Vertex Algebras
for Beginners

UNIVERSITY LECTURE SERIES VOLUME 10

Vertex Algebras for Beginners

Second Edition

Victor Kac

AMERICAN MATHEMATICAL SOCIETY

Providence, Rhode Island

2000 *Mathematics Subject Classification*. Primary 17B69;
Secondary 17B65, 81T05, 81T40.

ABSTRACT. This book is an introduction to vertex algebras, a new mathematical structure that has appeared recently in quantum physics. It can be used by researchers and graduate students working on representation theory and mathematical physics.

Library of Congress Cataloging-in-Publication Data

Kac, Victor G., 1943–
    Vertex algebras for beginners / Victor Kac. — 2nd ed.
      p. cm. — (University lecture series, ISSN 1047-3998 ; v. 10)
     Includes bibliographical references and index.
     ISBN 0-8218-1396-X (alk. paper)
     1. Vertex operator algebras. 2. Quantum field theory. 3. Mathematical physics. I. Title.
II. Series: University lecture series (Providence, R.I.) ; 10.
QC174.52.O6K33 1998
512′.55—dc21
                                                       98-41276
                                                            CIP

Contents

Preface

The notion of a vertex algebra was introduced ten years ago by Richard Borcherds [**B1**]. This is a rigorous mathematical definition of the chiral part of a 2-dimensional quantum field theory studied intensively by physicists since the landmark paper of Belavin, Polyakov and Zamolodchikov [**BPZ**]. However, implicitly this notion was known to physicists much earlier. Some of the most important precursors are Wightman axioms [**W**] and Wilson's notion of the operator product expansion [**Wi**]. In fact, as I show in Sections 1.1 and 1.2, the axioms of a vertex algebra can be deduced from Wightman axioms. The exposition of these two sections is somewhat terse. The rest of the book, written at a more relaxed pace, is motivated by these sections but can be read independently of them.

Axioms of a vertex algebra used in this book are essentially those of [**FKRW**] and were inspired by Goddard's lectures [**G**]. These axioms are much simpler than the original Borcherds' axioms and are very easy to check. One of the objectives of this book is to show that these systems of axioms are equivalent (see Section 4.8).

Another objective of the book is to lay rigorous grounds for the notion of the operator product expansion (OPE) and demonstrate how to use it to perform calculations that are otherwise very painful. The classical Wick theorem allows one to compute OPE in free field theories. A "non-commutative" generalization of Wick's formula allows one to compute OPE of arbitrary fields (see Section 3.3).

The main objective of the book is to show how to construct a variety of examples of vertex algebras, and how to perform calculations using the formalism of vertex algebras to get applications in many different directions (Chapter 5).

In Sections 2.7 and 5.10, I present some new material on a topic closely related to vertex algebras — the theory of conformal superalgebras.

These notes represent a part of the course given at MIT in 1994 and 1995. Unfortunately, I didn't have time to write down the chapters on representation theory of vertex algebras and some other applications. (Most quoted literature is related to these unwritten chapters, and I hope that the present book will facilitate the reading of these papers.) In fact, another important application of vertex algebra theory is that it picks out the most interesting representations of infinite-dimensional Lie (super)algebras and provides means for their detailed study.

There is nothing in this book on the application to the Monster simple group (there is a book [**FLM**] on this, after all), nothing on Borcherds' solution of the Conway-Norton problem [**B2**], and nothing on Borcherds' marvelous applications to generalized Kac-Moody algebras and automorphic forms [**B3**].

A technical remark is in order. What I call a "vertex algebra" should probably be called a "$N = 0$ vertex superalgebra" (see Section 5.9 for the definition of a $N = n$ vertex superalgebra), but I decided on this simpler name. (Also, I call a "conformal vertex algebra" what is called in [**FLM**], with some additional restrictions, a "vertex operator algebra.") The reader who detests "supermathematics" may assume that the $\mathbb{Z}_2$-gradation is trivial, that "Lie superalgebra" means "Lie algebra", etc. But then he skips fermions and beautiful applications to identities and to soliton equations, the rich variety of superconformal theories, etc.

The bibliography is by no means complete. It is already quite a task to compile a complete list that would include all the relevant work done by physicists. However, it includes all items that influenced my thinking on the subject. One may also find there further references.

In addition to the sources mentioned above, the most important for the present book were the work of Todorov on the Wightman axioms point of view on CFT, the paper by Li from which I learned the unified formula for n-th products and Dong's lemma, the paper by Getzler from which I learned the "non-commutative" Wick formula, and the work of Lian and Zuckerman on "quantum operator algebras."

A preliminary version of these notes has been published in the proceedings of the summer school in Bulgaria in 1995 where I lectured on this subject. I am grateful to Ivan Todorov and Kiyokazu Nagatomo for reading the manuscript and correcting errors, and to Maria Golenishcheva-Kutuzova, Mike Hopkins, Andrey Radul, and Ivan Todorov for numerous illuminating discussions.

Vienna, June 1996

Preface to the second edition

This improved and enlarged edition is based on a course given at M.I.T. in the spring of 1997 and in Rome University in May and June of 1997. Below is a list of the most important improvements and additions.

Chapter 2. The notion of formal Fourier transform is introduced in Section 2.2. This reduces significantly the calculations and leads to the important notion of λ-bracket in the theory of conformal algebras. Four new Sections 2.8–2.11 on the theory of conformal algebras are added and Section 2.7 is reworked. Thus, Sections 2.7–2.11 present the foundations of this rapidly developing area of algebraic conformal field theory.

Conformal algebra is an axiomatic description of the singular part of the operator product expansion of chiral fields in conformal field theory. It is, to some extent, related to a vertex algebra in the same way Lie algebra is related to its universal enveloping algebra. A structure theory of vertex algebras, similar, for example, to the structure theory of finite-dimensional Lie algebras, seems to be far away. Conformal algebras turned out to be a much more tractable object; as shown in Sections 2.7–2.11, for finite conformal algebras such a theory can be developed.

In Section 2.7 an explicit correspondence between an important class of infinite-dimensional Lie algebras, called formal distribution Lie algebras, and certain new structures, called conformal algebras, is established and a classification of finite conformal algebras is outlined. In Sections 2.8 and 2.9 representation theory of conformal algebras is developed, and in Section 2.11 the corresponding cohomology theory is explained. In Section 2.10 elements of conformal linear algebra are presented.

Chapter 3. The "non-commutative" Wick formula is expressed via λ-bracket (formula (3.3.12)), which greatly facilitates the use of this formula.

Chapter 4. The exposition of Sections 4.4-4.6 is simplified by making a more systematic use of the Uniqueness Theorem (a similar simplification was independently found in [**MN**]). Section 4.11 on field algebras is corrected.

Chapter 5. A new Section 5.8 on super boson-fermion correspondence is added. Comparing characters leads to a beautiful identity, whose specializations give classical results on sums of squares which go back to Gauss and Jacobi. In Section 5.10 a complete list of finite simple conformal superalgebras is given.

I wish to thank the participants of the course at M.I.T. for many discussions and suggestions, especially Bojko Bakalov, Alessandro D'Andrea, Eddie Karat, and Alexandre Soloviev. In particular, Bakalov gave a proof of Proposition 3.2 and suggested Example 4.11, and Karat and Soloviev gave proofs of Lemma 2.7. I am grateful to D. Fattori, A. Rudakov and J. van de Leur for sending corrections, and to Jan Wetzel for technical help in preparation of the manuscript. I am enormously indebted to Bojko Bakalov, Shun-Jen Cheng, Alessandro D'Andrea, Alexander Voronov, and Minoru Wakimoto for collaboration. It is due to their efforts that the theory of conformal algebras reached this level of maturity in such a short period of time.

Brookline, Massachusetts, December 1997

CHAPTER 1

Wightman axioms and vertex algebras

1.1. Wightman axioms of a QFT

Let M be the d-dimensional Minkowski space (space-time), i.e., the d-dimensional real vector space with metric

$$|x - y|^2 = (x_0 - y_0)^2 - (x_1 - y_1)^2 - \cdots - (x_{d-1} - y_{d-1})^2.$$

(As usual, $x_0 = ct$ where c is the speed of light and t is time, and $x_1, \ldots, x_{d-1}$ are space coordinates.)

Two subsets A and B of M are called space-like separated if for any $a \in A$ and $b \in B$ one has $|a - b|^2 < 0$. The forward cone is the set $\{x \in M \mid |x|^2 \geq 0,\ x_0 \geq 0\}$. Define causal order on M by $x \geq y$ iff $x - y$ lies in the forward cone.

The Poincaré group is the unity component of the group of all transformations of M preserving the metric. It is the semidirect product of the group of translations $(= M)$ and the Lorentz group L, the group of all unimodular linear transformations of M preserving the forward cone. Hence the Poincaré group preserves the causal order and therefore the space-like separateness.

A quantum field theory (QFT) is the following data:

> the *space of states*—a complex Hilbert space $\mathcal{H}$;
>
> the *vacuum vector*—a vector $|0\rangle \in \mathcal{H}$;
>
> a unitary representation $(q, \Lambda) \mapsto U(q, \Lambda)$ of the Poincaré group in $\mathcal{H}$;
>
> a collection of *fields* Φ_a (a an index)—operator-valued distributions on M (that is continuous linear functionals $f \mapsto \Phi_a(f)$ on the space of rapidly decreasing C^∞ tensor valued test functions on M with values in the space of linear operators densely defined on $\mathcal{H}$).

One requires that these data satisfy the following Wightman axioms:

W1 (Poincaré covariance): $U(q, \Lambda)\Phi_a(f)U(q, \Lambda)^{-1} = \Phi_a((q, \Lambda)f)), q \in M,$
$\Lambda \in L.$

Note that $U(q, 1) = \exp i \sum_{k=0}^{d-1} q_k P_k$, where P_k are self-adjoint commuting operators on $\mathcal{H}$.

> **W2 (stable vacuum):** The vacuum vector $|0\rangle$ is fixed by all the operators $U(q, \Lambda)$. The joint spectrum of all the operators $P_0, \ldots, P_{d-1}$ lies in the forward cone.
>
> **W3 (completeness):** The vacuum vector $|0\rangle$ is in the domain of any polynomial in the $\Phi_a(f)$'s and the linear subspace $\mathcal{D}$ of $\mathcal{H}$ spanned by all of them applied to $|0\rangle$ is dense in $\mathcal{H}$.
>
> **W4 (locality):** $\Phi_a(f)\Phi_b(h) = \Phi_b(h)\Phi_a(f)$ on $\mathcal{D}$ if the supports of f and h are spacelike separated.

The physical meaning of axoim W2 is that vacuum has zero energy and it is the minimal energy state. The last axiom means that the measurements in spacelike separated points are independent. (According to the main postulate of special relativity the speed of a signal does not exceed the speed of light.)

Actually, these are axioms of a purely "bosonic" QFT. In order to include "fermions" one considers even and odd fields by introducing parity $p(a) = \bar{0}$ or $\bar{1} \in \mathbb{Z}/2\mathbb{Z}$. Then only the axiom W4 is modified:

> **W4$_{\text{super}}$ (locality):** $\Phi_a(f)\Phi_b(h) = (-1)^{p(a)p(b)}\Phi_b(h)\Phi_a(f)$ on $\mathcal{D}$ if the supports of f and h are spacelike separated.

Axiom W1 gives, in particular, translation covariance ($q \in M$):

$$(1.1.1) \qquad\qquad U(q, 1)\Phi_a(x)U(q, 1)^{-1} = \Phi_a(x + q).$$

Here and further, by abuse of notation, we often write $\Phi_a(x)$ in place of $\Phi_a\big(f(x)\big)$.

Note that, by definition, $\mathcal{D}$ lies in the domain of definition and is invariant with respect to all the operators $\Phi_a(f)$. It follows from W1 and W2 that $\mathcal{D}$ is $U(q, 1)$-invariant. Since the translation covariance means

$$(1.1.2) \qquad\qquad i\,[P_k, \Phi_a] = \partial_{x_k}\Phi_a,$$

and $P_k|0\rangle = 0$ by W2, we see that $\mathcal{D}$ is invariant with respect to all the operators P_k.

Note that applying both sides of (1.1.1) to the vacuum vector and using its $U(q, 1)$-invariance, we obtain ($q \in M$):

$$(1.1.3) \qquad \Phi_a(x + q)|0\rangle = \left(\exp i \sum_k q_k P_k\right) \Phi_a(x)|0\rangle.$$

Now, the Poincaré group preserves distances on M. One considers also a larger group — the group of conformal transformations of M (preserving only angles). The simplest conformal transformation is the inversion

$$x \mapsto -x/|x|^2.$$

Conjugating a translation $x \mapsto x - b$ by the inversion, we get a special conformal transformation ($b \in M$):

$$(1.1.4) \qquad x^b = \frac{x + |x|^2 b}{1 + 2x \cdot b + |x|^2|b|^2}.$$

The group generated by the translations and the special conformal transformations is called the conformal group. It includes the Poincaré group and also the group of dilations:

$$x \mapsto \lambda x, \quad \lambda \neq 0.$$

Conformal transformations of the Minkowski space are important for QFT since they preserve causality (hence space-like separateness).

A quantum field theory is called *conformal* if the unitary representation of the Poincaré group in $\mathcal{H}$ extends to a unitary representation of the conformal group: $(q, \Lambda, b) \mapsto U(q, \Lambda, b)$ such that the vacuum vector $|0\rangle$ is still fixed and also the special conformal covariance holds for the given collection of fields; in the case of a scalar field it means

$$(1.1.5) \qquad U(0, 1, b)\Phi_a(x)\, U(0, 1, b)^{-1} = \varphi(b, x)^{-\Delta_a} \Phi_a\left(x^b\right),$$

where Δ_a is a real number called the *conformal weight* of the field Φ_a and

$$(1.1.6) \qquad \varphi(b, x) = 1 + 2x \cdot b + |x|^2|b|^2.$$

Note that $\varphi(b, x)^{-d}$ is the Jacobian of the transformation (1.1.4). It follows that axiom W1 and (1.1.5) together give conformal covariance:

$$U(q, \Lambda, b)\Phi_a(x)\, U(q, \Lambda, b)^{-1} = \varphi(b, x)^{-\Delta_a} \Phi_a((q, \Lambda, b) \cdot x).$$

In particular, we have dilation covariance:

$$(1.1.7) \qquad U(\lambda)\Phi_a(x)\,U(\lambda)^{-1} = \lambda^{\Delta_a}\Phi_a(\lambda x),$$

where $\lambda \mapsto U(\lambda)$ denotes the representation of the dilation subgroup.

Formula (1.1.5) implies that the infinitesimal special conformal generators are represented by selfadjoint operators Q_k $(k = 0, \ldots, d-1)$ on $\mathcal{H}$ such that

$$(1.1.8) \qquad i\,[Q_k, \Phi_a(x)] = \left(|x|^2 \partial_{x_k} - 2\eta_k x_k E - 2\Delta_a \eta_k x_k\right)\Phi_a(x),$$

where $E = \sum_{m=0}^{d-1} x_m \partial_{x_m}$ is the Euler operator and η_k are the coefficients of the metric ($\eta_0 = 1$, $\eta_k = -1$ for $k \geq 1$).

1.2. $d = 2$ QFT and chiral algebras

Consider now the case $d = 2$. Introduce the light cone coordinates $t = x_0 - x_1$, $\bar{t} = x_0 + x_1$, so that $|x|^2 = t\bar{t}$. (In this section the overbar does not mean the complex conjugate.) Let

$$P = \frac{1}{2}\left(P_0 - P_1\right), \qquad \bar{P} = \frac{1}{2}\left(P_0 + P_1\right).$$

Then formula (1.1.3) becomes:

$$(1.2.1) \qquad \Phi_a(t + q, \bar{t} + \bar{q})|0\rangle = e^{i(qP + \bar{q}\bar{P})}\Phi_a(t, \bar{t})|0\rangle.$$

By the vacuum axiom the joint spectrum of the operators P and $\bar{P}$ lies in the domain $t \geq 0$, $\bar{t} \geq 0$, hence the operator $\exp i(tP + \bar{t}\bar{P})$ is defined on $\mathcal{D}$ for all values $\operatorname{Im} t \geq 0$, $\operatorname{Im}\bar{t} \geq 0$. Moreover, by formula (1.2.1) the $\mathcal{D}$-valued distribution $\Phi_a|0\rangle$ extends analytically to a function in the domain

$$\{t \mid \operatorname{Im} t > 0\} \times \{\bar{t} \mid \operatorname{Im}\bar{t} > 0\} \subset \mathbb{C}^2.$$

Indeed, by the spectral decomposition, $e^{i(qP + \bar{q}\bar{P})}$ is the Fourier transform of a (operator valued) function whose support is in the domain $p \geq 0$, $\bar{p} \geq 0$, by the second part of axiom W2. Hence we may take the value $\Phi_a\,(t, \bar{t})\,|0\rangle$ when $\operatorname{Im} t > 0$, $\operatorname{Im}\bar{t} > 0$. It follows from (1.2.1) that this value is non-zero unless $\Phi_a = 0$.

The locality axiom means

$$(1.2.2) \quad \Phi_a\,(t, \bar{t})\,\Phi_b\,(t', \bar{t}') = (-1)^{p(a)p(b)}\Phi_b\,(t', \bar{t}')\,\Phi_a\,(t, \bar{t}) \quad \text{if} \quad (t - t')\,(\bar{t} - \bar{t}') < 0$$

In the light cone coordinates the special conformal transformations decouple:

$$(1.2.3) \qquad t^b = \frac{t}{1 + b_+ t}, \qquad \bar{t}^b = \frac{\bar{t}}{1 + b_- \bar{t}},$$

where $b_\pm = b_0 \pm b_1$. Hence the conformal group consists of transformations of the form:

$$\gamma(t, \bar{t}) = \left(\frac{at + b}{ct + d}, \frac{\bar{a}\bar{t} + \bar{b}}{\bar{c}\bar{t} + \bar{d}} \right),$$

where $\left(\begin{smallmatrix} a & d \\ c & b \end{smallmatrix} \right)$ and $\left(\begin{smallmatrix} \bar{a} & \bar{b} \\ \bar{c} & \bar{d} \end{smallmatrix} \right)$ are from $SL_2(\mathbb{R})$. Then the Poincaré covariance (axiom W1) and special conformal covariance (formula (1.1.5)) give together the following conformal covariance (with $\Delta_a = \bar{\Delta}_a$):

$$(1.2.4) \qquad U(\gamma)\Phi_a(t, \bar{t}) \, U(\gamma)^{-1} = (ct + d)^{-2\Delta_a} \left(\bar{c}\bar{t} + \bar{d} \right)^{-2\bar{\Delta}_a} \Phi_a(\gamma(t, \bar{t})).$$

Because of the decoupling (1.2.3) one usually does not assume that $\Delta_a = \bar{\Delta}_a$ and considers more general conformal covariance of the form (1.2.4).

Introduce further the operators

$$Q = -\frac{1}{2}(Q_0 + Q_1), \qquad \bar{Q} = \frac{1}{2}(Q_1 - Q_0).$$

Then formulas (1.1.2) and (1.1.8) become:

$$(1.2.5a) \qquad i\left[P, \Phi_a(t, \bar{t})\right] = \partial_t \Phi_a(t, \bar{t}),$$

$$(1.2.5b) \qquad i\left[\bar{P}, \Phi_a(t, \bar{t})\right] = \partial_{\bar{t}} \Phi_a(t, \bar{t}),$$

$$(1.2.5c) \qquad i\left[Q, \Phi_a(t, \bar{t})\right] = \left(t^2 \partial_t + 2\Delta_a t\right) \Phi_a(t, \bar{t}),$$

$$(1.2.5d) \qquad i\left[\bar{Q}, \Phi_a(t, \bar{t})\right] = \left(\bar{t}^2 \partial_{\bar{t}} + 2\bar{\Delta}_a \bar{t}\right) \Phi_a(t, \bar{t}).$$

In order to make conformal transformations defined everywhere, consider the compactification of the Minkowski space given by:

$$z = \frac{1 + it}{1 - it}, \qquad \bar{z} = \frac{1 + i\bar{t}}{1 - i\bar{t}}.$$

This maps the domain $\mathrm{Im}\, t > 0$, $\mathrm{Im}\, \bar{t} > 0$ to the domain $|z| < 1$, $|\bar{z}| < 1$. Consider the new fields defined in $|z| < 1$, $|\bar{z}| < 1$:

$$Y(a, z, \bar{z}) = \frac{1}{(1 + z)^{2\Delta_a}(1 + \bar{z})^{2\bar{\Delta}_a}} \Phi_a(t, \bar{t}), \qquad \text{where} \quad t = i\frac{1 - z}{1 + z}, \quad \bar{t} = i\frac{1 - \bar{z}}{1 + \bar{z}}.$$

Note that $Y(a, z, \bar{z})|0\rangle|_{z=0, \bar{z}=0}$ is a well defined vector in $\mathcal{D}$ which we denote by a, and (due to the above remark) $Y(a, z, \bar{z}) \mapsto a$ is a linear injective map.

We let

$$
\begin{aligned}
T &= \frac{1}{2}(P + [P,Q] - Q), \\
H &= \frac{1}{2}(P + Q), \\
T^* &= \frac{1}{2}(P - [P,Q] - Q),
\end{aligned}
$$

and similarly we define $\bar{T}$, $\bar{H}$, $\bar{T}^*$. It is straightforward to check that formulas (1.2.5a–d) imply:

$$
\text{(1.2.6a)} \qquad [T, Y(a,z,\bar{z})] = \partial_z Y(a,z,\bar{z}),
$$

$$
\text{(1.2.6b)} \qquad [H, Y(a,z,\bar{z})] = (z\partial_z + \Delta_a)Y(a,z,\bar{z}),
$$

$$
\text{(1.2.6c)} \qquad [T^*, Y(a,z,\bar{z})] = \left(z^2\partial_z + 2\Delta_a z\right) Y(a,z,\bar{z}),
$$

and similarly for $\bar{T}$, $\bar{H}$, $\bar{T}^*$. Also, of course, all the operators T, $\bar{T}$,... annihilate the vacuum vector $|0\rangle$.

Note that (1.2.6b) means:

$$
\lambda^H Y(a,z,\bar{z})\,\lambda^{-H} = \lambda^{\Delta_a} Y(a,\lambda z,\bar{z}).
$$

Note also that the operators T, H, and T^* satisfy the following commutation relations:

$$
\text{(1.2.7)} \qquad [H,T] = T, \quad [H,T^*] = -T^*, \quad [T^*,T] = 2H.
$$

Applying both sides of (1.2.6b and c) to the vacuum vector and letting $z = \bar{z} = 0$, we get:

$$
Ha = \Delta_a a, \quad T^* a = 0.
$$

Recall that P and $\bar{P}$ are positive semidefinite self-adjoint operators on $\mathcal{H}$ (due to axiom W2). The same is true for Q and $\bar{Q}$ since they are operators similar to P and $\bar{P}$ respectively. Hence H is a positive semidefinite self-adjoint operator as well. Thus, conformal weights are non-negative numbers.

If in our QFT, $Ta = 0 = \bar{T}a$ is possible only for the multiples of the vacuum vector, then $\Delta_a = \bar{\Delta}_a = 0$ imply that $a = |0\rangle$.

Now consider the right chiral fields, namely those fields for which $\partial_{\bar{t}} \Phi_a = 0$. Then (1.2.2) becomes

$$\Phi_a(t)\Phi_b(t') = (-1)^{p(a)p(b)}\Phi_b(t')\Phi_a(t) \quad \text{if} \quad t \neq t'.$$

This implies that the (super) commutator (i.e., the difference between the left- and the right-hand sides) has the following form:

$$[\Phi_a(t), \Phi_b(t')] = \sum_{j \geq 0} \delta^{(j)}(t - t')\Psi^j(t')$$

for some fields $\Psi^j(t')$. For these fields the Wightman axioms still hold (but the conformal covariance does not necessarily hold), hence we may add them to our QFT to obtain:

$$[Y(a, z), Y(b, w)] = \sum_{j \geq 0} \delta^{(j)}(z - w)Y(c_j, w).$$

Commuting H with both sides of this equality and using (1.2.6b) we see that the field $Y(c_j, w)$ has conformal weight $\Delta_a + \Delta_b - j - 1$ (in the sense of (1.2.6b)). Due to the positivity of conformal weights we conclude that the sum on the right is finite. It follows that

$$(z - w)^N[Y(a, z), Y(b, w)] = 0 \quad \text{for} \quad N \gg 0.$$

(A detailed explanation of this will be given in Section 2.3.)

We expand a chiral field $Y(a, z)$ in a Fourier series:

$$Y(a, z) = \sum_n a_{(n)} z^{-n-1},$$

where $a_{(n)} \in \text{End}\,\mathcal{D}$ and denote by V the subspace of $\mathcal{D}$ spanned by all polynomials in the $a_{(n)}$ applied to the vacuum vector $|0\rangle$. It is clear that V is invariant with respect to all $a_{(n)}$ and, by (1.2.6a), with respect to T. By the argument proving Corollary 4.6(f), V is spanned by all polynomials in the $a_{(n)}$ with $n < 0$ applied to $|0\rangle$.

We thus arrived at the following data called the right chiral algebra:

the space of states—a vector space V;

the vacuum vector—a non-zero vector $|0\rangle \in V$;

the infinitesimal translation operator $T \in \text{End}V$;

fields $Y(a, z)$ for each $a \in A$, some subset of V endowed with the parity $p(a)$, where

$$Y(a, z) = \sum_{n \in \mathbb{Z}} a_{(n)} z^{-n-1}$$

is a series with $a_{(n)} \in \mathrm{End} V$.

These data satisfy the following properties for $a \in A$ (we ignore the remaining properties for a while):

(translation covariance) $\quad [T, Y(a, z)] = \partial Y(z, a)$;

(vacuum) $\quad T|0\rangle = 0, \; Y(a, z)|0\rangle|_{z=0} = a$;

(completeness) $\quad$ polynomials in the $a_{(n)}$'s with $n < 0$ applied to $|0\rangle$ span V;

(locality) $\quad (z - w)^N Y(a, z) Y(b, w)$
$$= (-1)^{p(a)p(b)} (z - w)^N Y(b, w) Y(a, z) \text{ for some}$$
$N \in \mathbb{Z}_+$ (depending on $a, b \in A$).

By the vacuum property we have ($a \in A$):

$$(1.2.8) \qquad a = a_{(-1)}|0\rangle, \quad a_{(n)}|0\rangle = 0 \text{ for } n \geq 0.$$

Applying both sides of the translation covariance property to $|0\rangle$ and letting $z = 0$, we obtain (using $T|0\rangle = 0$ and (1.2.8)):

$$(1.2.9) \qquad Ta = a_{(-2)}|0\rangle, \quad a \in A.$$

Thus, the infinitesimal translation operator on A is built in the collection of fields.

The positivity of conformal weights imply, due to (1.2.6b):

$$(1.2.10) \qquad a_{(n)} v = 0 \text{ for } n \gg 0 (\text{depending on } a \in A \text{ and } v \in V).$$

Later (in Section 4.5) we shall prove the existence theorem that asserts that, using (1.2.10), one can construct fields $Y(a, z)$ for all $a \in V$ (using the so-called normally ordered product) such that (1.2.10), translation covariance, vacuum and locality properties still hold (completeness then automatically holds). We thus arrive at the definition of a chiral algebra. This name is used by physicists. Mathematicians, following Borcherds, use the name vertex algebras, or vertex operator algebras, since (for historical reasons) the fields $Y(a, z)$ are called vertex operators.

Similarly, one may consider the left chiral fields, that is those fields for which $\partial_l \Phi_{\bar{a}} = 0$. In the same way as above, we construct the left chiral algebra $\overline{V}$ with the same vacuum vector $|0\rangle$, the infinitesimal translation operator $\overline{T}$ and fields $Y(\bar{a}, z)$, $\bar{a} \in \overline{V}$. Due to locality (1.2.2) we see that $\Phi_a(t)\Phi_{\bar{a}}(\bar{t}) = (-1)^{p(a)p(\bar{a})}\Phi_{\bar{a}}(\bar{t})\Phi_a(t)$ for all t and $\bar{t}$, hence

$$[Y(a, z), Y(\bar{a}, \bar{z})] = 0 \text{ for all } a \in V, \quad \bar{a} \in \overline{V}.$$

The left and right chiral algebras are the most important invariants of a conformally covariant 2-dimensional QFT. Under certain assumptions and with certain additional data one may reconstruct the whole QFT from these chiral algebras, but we shall not discuss this problem here.

REMARK 1.2. One may also consider the case of $d = 1$ conformal QFT. Then the only coordinate is time $t = x_0$ and the forward cone is the set of non-negative numbers. Then conformal covariance reads:

$$U\left(\frac{at+b}{ct+d}\right) \Phi_a(t) \, U\left(\frac{at+b}{ct+d}\right)^{-1} = \frac{1}{(at+d)^{\Delta_a}} \Phi_a\left(\frac{at+b}{ct+d}\right).$$

It follows that there exist self-adjoint operators P and Q in $\mathcal{H}$ such that

$$i[P, \Phi_a(t)] = \partial_t \Phi_a(t), \quad i[Q, \Phi_a(t)] = \left(t^2\partial_t + 2\Delta_a t\right) \Phi_a(t).$$

Compactifying by $z = \frac{1+it}{1-it}$, letting

$$Y(a, z) = \frac{1}{(1+z)^{2\Delta_a}} \Phi_a(z)$$

and defining T, H, and T^* as in $d = 2$ case, we find that $Y(a, z)$ satisfies formulas (1.2.6a–c). As in $d = 2$ case, we see that $Y(a, z)|0\rangle \, |_{z=0}$ is a well-defined vector. The only property that is completely missing is locality since there are no spacelike separated points.

1.3. Definition of a vertex algebra

Let V be a *superspace*, i.e., a vector space decomposed in a direct sum of two subspaces:

$$V = V_{\bar{0}} + V_{\bar{1}}.$$

Here and further $\bar{0}$ and $\bar{1}$ stand for the cosets in $\mathbb{Z}/2\mathbb{Z}$ of 0 and 1. We shall say that an element a of V has *parity* $p(a) \in \mathbb{Z}/2\mathbb{Z}$ if $a \in V_{p(a)}$. If $\dim V$ $(= \dim V_{\bar{0}} + \dim V_{\bar{1}})$ $< \infty$, we let

$$\mathrm{sdim}\, V = \dim V_{\bar{0}} - \dim V_{\bar{1}}$$

to be the *superdimension* of V. In what follows, whenever $p(a)$ is written, it is to be understood that $a \in V_{p(a)}$.

A *field* is a series of the form $a(z) = \sum_{n \in \mathbb{Z}} a_{(n)} z^{-n-1}$ where $a_{(n)} \in \operatorname{End} V$ and for each $v \in V$ one has

$$(1.3.1) \qquad a_{(n)}(v) = 0 \quad \text{for} \quad n \gg 0.$$

We say that a field $a(z)$ has *parity* $p(a) \in \mathbb{Z}/2\mathbb{Z}$ if

$$(1.3.2) \qquad a_{(n)} V_\alpha \subset V_{\alpha + p(a)} \quad \text{for all} \quad \alpha \in \mathbb{Z}/2\mathbb{Z}, \quad n \in \mathbb{Z}.$$

A *vertex algebra* is the following data:

> the space of states—a superspace V,
>
> the vacuum vector—a vector $|0\rangle \in V_{\bar{0}}$,
>
> the state-field correspondence—a parity preserving linear map
> of V to the space of fields, $a \mapsto Y(a, z) = \sum_{n \in \mathbb{Z}} a_{(n)} z^{-n-1}$,
> satisfying the following axioms:
>
> **(translation covariance):** $[T, Y(a, z)] = \partial Y(a, z),$
>
> > where $T \in \operatorname{End} V$ is defined by

$$(1.3.3) \qquad T(a) = a_{(-2)} |0\rangle,$$

> **(vacuum):** $Y(|0\rangle, z) = I_V$, $Y(a, z)|0\rangle|_{z=0} = a,$
>
> **(locality):** $(z - w)^N Y(a, z) Y(b, w)$
> $= (-1)^{p(a)p(b)} (z - w)^N Y(b, w) Y(a, z)$ for $N \gg 0.$

Note that the *infinitesimal translation operator T* is an even operator, i.e., $T V_\alpha \subset V_\alpha$, and the bracket in the translation covariance axiom is the usual bracket: $[T, Y] = TY - YT$, so that this axiom says

$$(1.3.4) \qquad \left[T, a_{(n)} \right] = -n a_{(n-1)}.$$

The first of the vacuum axioms says that

$$(1.3.5a) \qquad |0\rangle_{(n)} = \delta_{n,-1}; \text{ in particular } T|0\rangle = 0.$$

The second of the vacuum axioms says that

$$(1.3.5b) \qquad a_{(n)}|0\rangle = 0 \text{ for } n \geq 0, \quad a_{(-1)}|0\rangle = a.$$

The locality axiom is to be understood as a coefficient-wise equality of two series in z and w of the form $\sum_{m,n\in\mathbb{Z}} a_{m,n} z^m w^n$.

REMARK 1.3. Applying T to both sides of (1.3.3) $n-1$ times, and using (1.3.4) and $T|0\rangle = 0$, we obtain $\frac{T^n}{n!}(a) = a_{(-n-1)}|0\rangle$, for $n \in \mathbb{Z}_+$, which is equivalent, by (1.3.5b), to

$$(1.3.6) \qquad Y(a,z)|0\rangle = e^{zT}(a).$$

1.4. Holomorphic vertex algebras

A vertex algebra V is called *holomorphic* if $a_{(n)} = 0$ for $n \geq 0$, i.e., $Y(a,z) = \sum_{n\in\mathbb{Z}_+} a_{(-n-1)} z^n$ are formal power series in z.

Let V be a holomorphic vertex algebra. Since the algebra of formal power series in z and w has no zero divisors, it follows that locality for V turns into a usual supercommutativity:

$$(1.4.1) \qquad Y(a,z)Y(b,w) = (-1)^{p(a)p(b)} Y(b,w)Y(a,z).$$

Define a bilinear product ab on the space V by the formula

$$(1.4.2) \qquad ab = a_{(-1)}b$$

and let $|0\rangle = 1$. Then applying both sides of (1.4.1) to c and letting $z = w = 0$ gives:

$$(1.4.3) \qquad a(bc) = (-1)^{p(a)p(b)} b(ac).$$

The vacuum axioms give

$$(1.4.4) \qquad 1 \cdot a = a \cdot 1 = a.$$

It is easy to see that properties (1.4.3) and (1.4.4) are equivalent to the axioms of a (super)commutative associative unital super algebra. Indeed, letting $c = 1$ in (1.4.3), we see by (1.4.4) that V is (super)commutative. But using (super)commutativity, we can rearrange (1.4.3) to get $a(cb) = (ac)b$, which is associativity. The converse is clear.

Furthermore, apply $Y(b,w)$ to both sides of (1.3.6):

$$Y(b,w)Y(a,z)1 = Y(b,w)e^{zT}(a).$$

Applying commutativity to the left-hand side and then (1.3.6), we obtain

$$(-1)^{p(a)p(b)} Y(a,z) e^{wT}(b) = Y(b,w) e^{zT}(a).$$

Letting $w = 0$ and using the commutativity of our product on V we get

$$(1.4.5) \qquad Y(a,z)(b) = e^{zT}(a)b.$$

Thus, the fields $Y(a,z)$ are defined entirely in terms of the product on V and the operator T.

Finally, by (1.4.5), translation covariance axiom becomes:

$$(1.4.6) \qquad T\left(e^{zT}(a)b\right) - e^{zT}(a)T(b) = T\left(e^{zT}(a)\right)b.$$

Letting $z = 0$ we see that T is an even derivation of the associative commutative unital superalgebra V and that (1.4.6) is equivalent to this.

Thus, we canonically associated to a holomorphic vertex algebra V a pair consisting of an associative commutative unital superalgebra structure on V and an even derivation T. Conversely, to such a pair we canonically associate a holomorphic vertex algebra with fields defined by (1.4.5).

If $T = 0$, then $Y(a,z)(b) = ab$. Therefore we may view vertex algebras as a generalization of unital commutative associative superalgebras where the multiplication depends on the parameter z via

$$a_z b = Y(a,z)(b).$$

However, as we shall see, a general vertex algebra is very far from being a "commutative" object.

CHAPTER 2

Calculus of formal distributions

2.1. Formal delta-function

In the previous chapter we considered formal expressions

$$(2.1.1) \qquad \sum_{m,n,\ldots\in\mathbb{Z}} a_{m,n,\ldots} z^m w^n \ldots ,$$

where $a_{m,n,\ldots}$ are elements of a vector space U over $\mathbb{C}$. Series of the form (2.1.1) are called *formal distributions* in the indeterminates $z, w, \ldots$ with values in U. They form a vector space over $\mathbb{C}$ denoted by $U\left[\left[z, z^{-1}, w, w^{-1}, \ldots\right]\right]$.

We can always multiply a formal distribution and a Laurent polynomial (provided that product of coefficients is defined), but cannot in general multiply two formal distributions. Each time when a product of two formal distribution occurs, we need to check that it converges in the algebraic sense, i.e. the coefficient of each monomial $z^m w^n \ldots$ is a finite (or convergent) sum.

Given a formal distribution $a(z) = \sum_{n\in\mathbb{Z}} a_n z^n$, we define the *residue* by the usual formula

$$\mathrm{Res}_z\, a(z) = a_{-1}.$$

Since $\mathrm{Res}_z\, \partial a(z) = 0$, we have the usual integration by parts formula (provided that ab is defined):

$$(2.1.2) \qquad \mathrm{Res}_z\, \partial a(z)b(z) = -\,\mathrm{Res}_z\, a(z)\partial b(z).$$

Here and further $\partial a(z) = \sum_n n a_n z^{n-1}$ is the derivative of $a(z)$.

Let $\mathbb{C}\left[z, z^{-1}\right]$ denote the algebra of Laurent polynomials in z. We have a non-degenerate pairing $U\left[\left[z, z^{-1}\right]\right] \times \mathbb{C}\left[z, z^{-1}\right] \to U$ defined by $\langle f, \varphi\rangle = \mathrm{Res}_z\, f(z)\varphi(z)$, hence the Laurent polynomials should be viewed as "test functions" for the formal distributions. Note that formal distributions $a(z)$ and $b(z)$ are equal iff $\langle a, \varphi\rangle = \langle b, \varphi\rangle$ for any test function $\varphi \in \mathbb{C}\left[z, z^{-1}\right]$.

17

We introduce the *formal delta-function* $\delta(z - w)$ as the following formal distribution in z and w with values in $\mathbb{C}^1$:

$$(2.1.3) \qquad \delta(z - w) = z^{-1} \sum_{n \in \mathbb{Z}} \left(\frac{w}{z}\right)^n.$$

In order to establish its properties, introduce one more notation. Given a rational function $R(z, w)$ with poles only at $z = 0$, $w = 0$ and $|z| = |w|$, we denote by $i_{z,w} R$ (resp. $i_{w,z} R$) the power series expansion of R in the domain $|z| > |w|$ (resp. $|w| > |z|$). For example, we have for $j \in \mathbb{Z}_+$:

$$(2.1.4a) \qquad i_{z,w} \frac{1}{(z - w)^{j+1}} = \sum_{m=0}^{\infty} \binom{m}{j} z^{-m-1} w^{m-j},$$

$$(2.1.4b) \qquad i_{w,z} \frac{1}{(z - w)^{j+1}} = -\sum_{m=-1}^{-\infty} \binom{m}{j} z^{-m-1} w^{m-j}.$$

From (2.1.3) and (2.1.4a and b) we obtain the following important formula:

$$(2.1.5a) \qquad \partial_w^{(j)} \delta(z - w) = i_{z,w} \frac{1}{(z - w)^{j+1}} - i_{w,z} \frac{1}{(z - w)^{j+1}}$$

$$(2.1.5b) \qquad = \sum_{m \in \mathbb{Z}} \binom{m}{j} z^{-m-1} w^{m-j}.$$

Here and further for an operator A we let

$$(2.1.6) \qquad A^{(j)} = A^j / j!.$$

Note that (2.1.5a) is a formal distribution with integer coefficients.

The formal delta-function has the usual properties listed below.

PROPOSITION 2.1. (a) *For any formal distribution $f(z) \in U\left[\left[z, z^{-1}\right]\right]$ one has:*

$$(2.1.7) \qquad \operatorname{Res}_z f(z) \delta(z - w) = f(w).$$

(The product $f(z)\delta(z - w)$ always converges.)

(b) $\delta(z - w) = \delta(w - z)$.

(c) $\partial_z \delta(z - w) = -\partial_w \delta(z - w)$.

(d) $(z - w)\partial_w^{(j+1)} \delta(z - w) = \partial_w^{(j)} \delta(z - w)$, $j \in \mathbb{Z}_+$.

(e) $(z - w)^{j+1} \partial_w^{(j)} \delta(z - w) = 0$, $j \in \mathbb{Z}_+$.

---

[1]This notation is very suggestive but somewhat misleading as $\delta(z - w)$ is not a function of $z - w$. Purists may use notation $\delta(z, w)$.

PROOF. It suffices to check (2.1.7) for $f(z) = az^n$, which is straightforward. Furthermore, we have:

$$\delta(z - w) = z^{-1} \sum_{n \in \mathbb{Z}} \left(\frac{w}{z}\right)^{n-1} = w^{-1} \sum_{n \in \mathbb{Z}} \left(\frac{w}{z}\right)^{n} = \delta(w - z),$$

proving (b). Since $\delta(z - w) = \sum_m z^{-m-1} w^m = \sum_m z^{-m-2} w^{m+1}$, we see that $\partial_z \delta(z - w) = -\partial_w \delta(z - w)$, proving (c). Finally, (d) and (e) follow from (2.1.5a and b). $\qquad\square$

Note that Proposition 2.1 (c-e) can be also proved by comparing the values of both sides on test functions. Let us use this method in order to prove the following useful formula (which is a generalization of Proposition 2.1 (e) for $j = 0$):

$$(2.1.8) \qquad \delta(z - w)a(z) = \delta(z - w)a(w), \quad \text{where} \quad a(z) \in U\left[\left[z, z^{-1}\right]\right].$$

Indeed, by (2.1.7), the pairing of both sides of (2.1.8) with $\varphi(z) \in \mathbb{C}\left[z, z^{-1}\right]$ is equal to $a(w)\varphi(w)$.

Letting $a(z) = \delta(z - t)$, we obtain an important special case of (2.1.8), after exchanging t and z:

$$(2.1.9) \qquad \delta(z - t)\delta(w - t) = \delta(w - t)\delta(z - w).$$

Applying to both sides $\partial_z^n \partial_w^m$ and using Proposition 2.1 (c) we obtain

$$(2.1.10) \qquad \partial_t^n \delta(z - t)(-\partial_t)^m \delta(w - t) = \sum_{j=0}^{m} \binom{m}{j} \partial_w^{m-j} \delta(w - t) \partial_w^{n+j} \delta(z - w).$$

This formula is very useful for checking locality of formal distributions.

2.2. An expansion of a formal distribution $a(z, w)$ and formal Fourier transform

Here we consider the question: when a formal distribution

$$a(z, w) = \sum_{m,n \in \mathbb{Z}} a_{m,n} z^m w^n \in U\left[\left[z, z^{-1}, w, w^{-1}\right]\right]$$

has an expansion of the form

$$(2.2.1) \qquad a(z, w) = \sum_{j=0}^{\infty} c^j(w) \partial_w^{(j)} \delta(z - w).$$

Multiplying both sides of (2.2.1) by $(z - w)^n$ and taking Res_z we obtain using Proposition 2.1 (a, d, and e)

$$(2.2.2) \qquad c^n(w) = \mathrm{Res}_z\, a(z, w)(z - w)^n.$$

Denote by $U\left[\left[z, z^{-1}, w, w^{-1}\right]\right]^0$ the subspace consisting of formal U-valued distributions $a(z, w)$ for which the following series converges:

$$(2.2.3) \qquad \pi a(z, w) := \sum_{j=0}^{\infty} \left(\mathrm{Res}_z\, a(z, w)(z - w)^j\right) \partial_w^{(j)} \delta(z - w).$$

Let

$$(2.2.4) \qquad a(z, w)^{+(z)} := \sum_{\substack{m \in \mathbb{Z}_+ \\ n \in \mathbb{Z}}} a_{m,n} z^m w^n.$$

A formal distribution $a(z, w)$ is called *holomorphic in z* if $a(z, w) = a(z, w)^{+(z)}$.

PROPOSITION 2.2. (a) *The map π is a projector (i.e., $\pi^2 = \pi$) on* $U\left[\left[z, z^{-1}, w, w^{-1}\right]\right]^0$.

(b) $\mathrm{Ker}\,\pi = \left\{a(z, w) \in U\left[\left[z, z^{-1}, w, w^{-1}\right]\right]^0 \text{ which are holomorphic in } z\right\}$.

(c) *Any formal distribution $a(z, w)$ from* $U\left[\left[z, z^{-1}, w, w^{-1}\right]\right]^0$ *is uniquely represented in the form:*

$$(2.2.5) \qquad a(z, w) = \sum_{j=0}^{\infty} c^j(w) \partial_w^{(j)} \delta(z - w) + b(z, w)$$

where $b(z, w)$ is a formal distribution holomorphic in z. The coefficients $c^j(w)$ are given by (2.2.2).

PROOF. (a) follows by the argument preceding formula (2.2.2). It is clear that $a(z, w) \in \mathrm{Ker}\,\pi$ if $a(z, w)$ is holomorphic in z. Conversely, if $a(z, w) \in \mathrm{Ker}\,\pi$, writing $a(z, w) = \sum_{n \in \mathbb{Z}} a_n(w) z^n$, we see from (2.2.2) that $c^0(w) = 0$ implies $a_{-1}(w) = 0$, $c^0(w) = c^1(w) = 0$ implies $a_{-1}(w) = a_{-2}(w) = 0$, etc., proving (b). (c) follows from (a) and (b). $\qquad \square$

COROLLARY 2.2. *The null space of the operator of multiplication by $(z - w)^N$, $N \geq 1$, in* $U\left[\left[z, z^{-1}, w, w^{-1}\right]\right]$ *is*

$$(2.2.6) \qquad \sum_{j=0}^{N-1} \partial_w^{(j)} \delta(z - w) U\left[\left[w, w^{-1}\right]\right].$$

Any element $a(z, w)$ from (2.2.6) is uniquely represented in the form

$$(2.2.7) \qquad a(z, w) = \sum_{j=0}^{N-1} c^j(w) \partial_w^{(j)} \delta(z - w),$$

the $c^j(w)$ being given by (2.2.2).

PROOF. That (2.2.6) lies in the null space of $(z - w)^N$ follows from Proposition 2.1e.

Conversely, if $(z - w)^N a(z, w) = 0$, then $a(z, w) \in U\left[\left[z, z^{-1}, w, w^{-1}\right]\right]^0$ and we have by (2.2.5) and Proposition 2.1 (d and e):

$$0 = \sum_{j=0}^{\infty} c^{j+N}(w) \partial_w^{(j)} \delta(z - w) + (z - w)^N b(z, w).$$

By the uniqueness in Proposition 2.2c we conclude that $c^j(w) = 0$ for $j \geq N$ and that $(z - w)^N b(z, w) = 0$. The last equality implies $b(z, w) = 0$ since $b(z, w) = \sum_{n \in \mathbb{Z}_+} a_n(w) z^n$. $\qquad \square$

We shall often write a formal distribution in the form

$$a(z) = \sum_{n \in \mathbb{Z}} a_{(n)} z^{-n-1}, \quad a(z, w) = \sum_{m,n \in \mathbb{Z}} a_{(m,n)} z^{-m-1} w^{-n-1}, \text{etc.}$$

This is a natural thing to do since $a_{(n)} = \operatorname{Res}_z a(z) z^n$. Then the expansion (2.2.7) is equivalent to

$$(2.2.8) \qquad a_{(m,n)} = \sum_{j=0}^{N-1} \binom{m}{j} c^j_{(m+n-j)}.$$

This follows by using (2.1.5b) and comparing coefficients.

DEFINITION 2.2. A formal distribution $a(z, w)$ is called *local* if

$$(z - w)^N a(z, w) = 0 \quad \text{for} \quad N \gg 0.$$

Corollary 2.2 says that any local formal distribution $a(z, w)$ has the expansion (2.2.7). This expansion is called the *OPE expansion* of $a(z, w)$ and the $c^n(w)$ (given by (2.2.2)) are called the *OPE coefficients* of $a(z, w)$.

In order to study the properties of the expansion (2.2.5), it is convenient to introduce the *formal Fourier transform* of a formal distribution $a(z, w)$ by the formula:

$$F_{z,w}^{\lambda}(a(z, w)) = \operatorname{Res}_z e^{\lambda(z-w)} a(z, w).$$

This is a $\mathbb{C}$-linear map from $U\left[\left[z, z^{-1}, w, w^{-1}\right]\right]$ to $U\left[\left[w, w^{-1}\right]\right]\left[\left[\lambda\right]\right]$. It is immediate, by Proposition 2.1(d and e) and (2.1.7), that

$$F_{z,w}^{\lambda}\left(\partial_w^j \delta(z-w)\right) = \lambda^j.$$

Hence the formal Fourier transform of the expansion (2.2.5) is

$$(2.2.9) \qquad F_{z,w}^{\lambda}(a(z,w)) = \sum_{n \in \mathbb{Z}_+} \lambda^{(n)} c^n(w).$$

(As before, $\lambda^{(n)}$ stands for $\lambda^n/n!$.) In other words, the formal Fourier transform of a formal distribution $a(z,w)$ is the generating series of its OPE coefficients.

The following simple lemma is very useful.

LEMMA 2.2.

$$e^{\lambda(z-w)} \partial_w^j \delta(z-w) = (\lambda + \partial_w)^j \delta(z-w).$$

PROOF. It is straightforward using Proposition 2.1 (d) and (e).   □

Along with the operators ∂_z and ∂_w on the space of formal distribution $U\left[\left[z, z^{-1}, w, w^{-1}\right]\right]$, consider the permutation operator $\tilde{a}(z,w) = a(w,z)$. It is clear that all three operators preserve the property of locality. The following formulas describe the behavior of the formal Fourier transform with respect to these operators:

$$(2.2.10) \qquad F_{z,w}^{\lambda} \partial_z = -\lambda F_{z,w}^{\lambda} = \left[\partial_w, F_{z,w}^{\lambda}\right],$$

$$(2.2.11) \qquad F_{z,w}^{\lambda} a(w,z) = F_{z,w}^{-\lambda-\partial_w} a(z,w) \quad \text{if } a(z,w) \text{ is local.}$$

(The right-hand side of (2.2.11) means that the indeterminate λ in (2.2.9) is replaced by the operator $-\lambda - \partial_w$.) Formulas (2.2.10) follow from the definition of $F_{z,w}^{\lambda}$ using integration by parts (they hold without the assumption of locality). Due to locality of $a(z,w)$, we can use expansion (2.2.7), hence it suffices to check (2.2.11) for $a(z,w) = c(w)\partial_w^k \delta(z-w)$. We use Proposition 2.1(c) and Lemma 2.2:

$$\begin{aligned}
F_{z,w}^{\lambda} a(w,z) &= (-1)^k \operatorname{Res}_z \left(e^{\lambda(z-w)} c(z)\partial_w^k \delta(z-w)\right) \\
&= (-1)^k (\lambda + \partial_w)^k \operatorname{Res}_z c(z)\delta(z-w) \\
&= (-\lambda - \partial_w)^k c(w).
\end{aligned}$$

REMARK 2.2a. Formulas (2.2.10) and (2.2.11) are equivalent to the following relations for the OPE coefficients $c_z^n(w)$, $c_w^n(w)$ and $\tilde{c}^n(w)$ of the formal distributions $\partial_z a(z,w)$, $\partial_w a(z,w)$ and $\tilde{a}(z,w)$ respectively:

$$c_z^n(w) = -nc^{n-1}(w),$$
$$c_w^n(w) = \partial_w c^n(w) + nc^{n-1}(w),$$
$$\tilde{c}^n(w) = \sum_{j \in \mathbb{Z}_+} (-1)^{j+n} \partial_w^{(j)} c^{n+j}(w).$$

A composition of two Fourier transforms, $F_{z,w}^\lambda F_{x,w}^\mu$ is a $\mathbb{C}$-linear map from $U\left[\left[z, z^{-1}, w, w^{-1}, x, x^{-1}\right]\right]$ to $U\left[\left[w, w^{-1}\right]\right][[\lambda, \mu]]$. The following relation will simplify significantly our calculations:

$$(2.2.12) \qquad F_{z,w}^\lambda F_{x,w}^\mu = F_{x,w}^{\lambda+\mu} F_{z,x}^\lambda.$$

The proof of it is very easy. Indeed, the left-hand side applied to $a(z,w,x)$ is equal to $\mathrm{Res}_z \mathrm{Res}_x e^{\lambda(z-w)+\mu(x-w)} a(z,w,x) = \mathrm{Res}_x \mathrm{Res}_z e^{\lambda(z-x)} e^{(\lambda+\mu)(x-w)} a(z,w,x)$, which is the right-hand side applied to $a(z,w,x)$.

REMARK 2.2b. A language alternative to that of U-valued local formal distributions in z and w is the language of differential operators from $U\left[w, w^{-1}\right]$ to $U\left[\left[w, w^{-1}\right]\right]$. Indeed, for a formal distribution $a(z,w)$ the associated operator is

$$\left(D_{a(z,w)} f\right)(w) = \mathrm{Res}_z a(z,w) f(z).$$

It is easy to see that $D_{\partial_w^k \delta(z-w)} = \partial_w^k$ ($k \in \mathbb{Z}_+$), hence for

$$a(z,w) = \sum_k c^k(w) \partial_w^{(k)} \delta(z-w),$$

we have:

$$D_{a(z,w)} = \sum_k c^k(w) \partial_w^{(k)}.$$

Note that we also have:

$$D_{a(w,z)} = \sum_k (-\partial_w)^{(k)} c^k(w).$$

2.3. Locality of two formal distributions

Suppose now that the vector space U carries a structure of an *associative superalgebra*. This simply means that $U = U_{\bar{0}} \oplus U_{\bar{1}}$ is a $\mathbb{Z}/2\mathbb{Z}$-graded associative algebra (i.e., $U_\alpha U_\beta \subset U_{\alpha+\beta}$, $\alpha, \beta \in \mathbb{Z}/2\mathbb{Z}$).

The most important example of an associative superalgebra is the endomorphism algebra $\mathrm{End}V$ of a superspace V (see Section 1.3) with the $\mathbb{Z}/2\mathbb{Z}$-grading given by:

$$(\mathrm{End}V)_\alpha = \{a \in \mathrm{End}V \mid aV_\beta \subset V_{\alpha+\beta}\}.$$

One defines the *bracket* $[\,,\,]$ on an associative superalgebra U by letting

$$(2.3.1) \qquad [a,b] = ab - p(a,b)ba, \quad \text{where} \quad a \in U_\alpha,\ b \in U_\beta,\ p(a,b) = (-1)^{\alpha\beta}.$$

Here and further we adopt the convention of [**K1**] that the bracket of an even element with any other element is the usual commutator and the bracket of two odd elements is the anti-commutator (physicists usually write $[a,b]_+$ in the latter case). Recall that the $\mathbb{Z}/2\mathbb{Z}$-graded space U with the bracket $(2.3.1)$ is a basic example of a Lie superalgebra (see e.g. [**K1**] for a definition).

We can define now the notion of locality of formal distributions, with values in a Lie superalgebra $\mathfrak{g}$, hence in its universal enveloping algebra $U(\mathfrak{g})$.

DEFINITION 2.3. Two formal distributions $a(z)$ and $b(z)$ with values in a Lie superalgebra $\mathfrak{g}$ are called *mutually local* (or *simply* local, or form a *local pair*) if the formal distribution $[a(z), b(w)] \in \mathfrak{g}\left[\left[z, z^{-1}, w, w^{-1}\right]\right]$ is local, i.e. if

$$(2.3.2) \qquad (z-w)^N [a(z), b(w)] = 0 \quad \text{for} \quad N \gg 0.$$

We shall always assume that all coefficients of a formal distribution $a(z)$ have the same parity, which will be denoted by $p(a)$. We shall also use the following notation:

$$p(a,b) = (-1)^{p(a)p(b)}.$$

REMARK 2.3a. Differentiating both sides of $(2.3.2)$ by z and multiplying by $z-w$, we see that the locality of $a(z)$ and $b(z)$ implies the locality of $\partial a(z)$ and $b(z)$.

In order to state equivalent definitions of locality we need some notation. Given a formal distribution $a(z) = \sum_{n \in \mathbb{Z}} a_{(n)} z^{-n-1}$, let

$$(2.3.3) \qquad a(z)_- = \sum_{n \geq 0} a_{(n)} z^{-n-1}, \quad a(z)_+ = \sum_{n < 0} a_{(n)} z^{-n-1}.$$

This is the only way to break $a(z)$ into a sum of "positive" and "negative" parts such that

$$(2.3.4) \qquad (\partial a(z))_\pm = \partial(a(z)_\pm).$$

Given formal distributions $a(z)$ and $b(z)$, define the following formal distribution in z and w with values in the universal enveloping algebra $U(\mathfrak{g})$:

$$(2.3.5) \qquad : a(z)b(w) := a(z)_+ b(w) + p(a,b)b(w)a(z)_-.$$

Note the following formulas:

$$(2.3.6a) \qquad a(z)b(w) \;=\; [a(z)_-, b(w)] + : a(z)b(w) :$$

$$(2.3.6b) \qquad p(a,b)b(w)a(z) \;=\; -[a(z)_+, b(w)] + : a(z)b(w) :$$

THEOREM 2.3. *Each of the following properties* (i)–(vii) *is equivalent to* (2.3.2):

(i) $[a(z), b(w)] = \sum\limits_{j=0}^{N-1} \partial_w^{(j)} \delta(z-w) c^j(w)$, *where* $c^j(w) \in \mathfrak{g}\left[\left[w, w^{-1}\right]\right]$.

(ii) $[a(z)_-, b(w)] = \sum\limits_{j=0}^{N-1} \left(i_{z,w} \dfrac{1}{(z-w)^{j+1}} \right) c^j(w),$

$\quad -[a(z)_+, b(w)] = \sum\limits_{j=0}^{N-1} \left(i_{w,z} \dfrac{1}{(z-w)^{j+1}} \right) c^j(w),$

$\quad$ *where* $c^j(w) \in \mathfrak{g}\left[\left[w, w^{-1}\right]\right]$.

(iii) $a(z)b(w) = \sum\limits_{j=0}^{N-1} \left(i_{z,w} \dfrac{1}{(z-w)^{j+1}} \right) c^j(w) + : a(z)b(w) :,$

$\quad p(a,b)b(w)a(z) = \sum\limits_{j=0}^{N-1} \left(i_{w,z} \dfrac{1}{(z-w)^{j+1}} \right) c^j(w) + : a(z)b(w) :,$

$\quad$ *where* $c^j(w) \in \mathfrak{g}\left[\left[w, w^{-1}\right]\right]$.

(iv) $[a_{(m)}, b_{(n)}] = \sum\limits_{j=0}^{N-1} \binom{m}{j} c^j_{(m+n-j)}, \quad m, n \in \mathbb{Z}.$

(v) $[a_{(m)}, b(w)] = \sum\limits_{j=0}^{N-1} \binom{m}{j} c^j(w) w^{m-j}, \quad m \in \mathbb{Z}.$

$$\text{(vi)} \quad \left[a_{(m)}, b_{(n)}\right] = \sum_{j=0}^{N-1} p_j(m) d_{m+n}^j, \quad m, n \in \mathbb{Z},$$

for some polynomials $p_j(x)$ and elements d_k^j of $\mathfrak{g}$.

$$\text{(vii)} \qquad a(z)b(w) \;=\; \left(i_{z,w} \frac{1}{(z-w)^N}\right) c(z,w),$$

$$p(a,b)b(w)a(z) \;=\; \left(i_{w,z} \frac{1}{(z-w)^N}\right) c(z,w)$$

for a formal distribution $c(z,w)$.

PROOF. (i) is equivalent to (2.3.2) due to Corollary 2.2. (ii) is equivalent to (i) by taking all terms in (i) with negative (resp. non-negative) powers of z. (iii) is equivalent to (ii) due to ((2.3.6a) and b). (iv) and (v) are equivalent to (i) due to (2.2.8). (vi) is equivalent to (iv) since any polynomial is a linear combination of binomial coefficients. Finally, (iii) implies (vii) and (vii) implies (2.3.2). $\square$

By abuse of notation physicists write the first of the relations of Theorem 2.3(iii) as follows:

$$\text{(2.3.7a)} \qquad a(z)b(w) = \sum_{j=0}^{N-1} \frac{c^j(w)}{(z-w)^{j+1}} + : a(z)b(w) :,$$

or often write just the singular part:

$$\text{(2.3.7b)} \qquad a(z)b(w) \sim \sum_{j=0}^{N-1} \frac{c^j(w)}{(z-w)^{j+1}}.$$

Formulas (2.3.7a) and (2.3.7b) are called the *operator product expansion* (OPE). By Theorem 2.3 the singular part of the OPE encodes all the brackets between all the coefficients of mutually local formal distributions $a(z)$ and $b(z)$. That is why it is important to develop techniques for the calculation of the OPE's. Most of the time we shall use the form (2.3.7b) of the OPE as typographically the most convenient.

For each $n \in \mathbb{Z}_+$ introduce the *n-th product* $a(w)_{(n)}b(w)$ on the space of formal distributions by the formula

$$\text{(2.3.8)} \qquad a(w)_{(n)}b(w) = \mathrm{Res}_z \, [a(z), b(w)] \, (z-w)^n.$$

Then, due to Corollary 2.2, the OPE (2.3.7a) becomes (for any two local formal distributions $a(z)$ and $b(z)$):

$$\text{(2.3.9a)} \qquad a(z)b(w) = \sum_{j=0}^{N-1} \frac{a(w)_{(j)}b(w)}{(z-w)^{j+1}} + : a(z)b(w) : .$$

Equivalently:

$$(2.3.9b) \qquad [a(z), b(w)] = \sum_{j \in \mathbb{Z}_+} \left(a(w)_{(j)} b(w)\right) \partial_w^{(j)} \delta(z - w).$$

As we have seen in the previous section, an efficient way to study the OPE is to consider its formal Fourier transform.

For an arbitrary (not necessarily associative or Lie) algebra U define the λ-product $a(w)_\lambda b(w)$ of two U-valued formal distributions $a(w)$ and $b(w)$ as the formal Fourier transform of the formal distribution $a(z)b(w)$:

$$(2.3.10a) \qquad a(w)_\lambda b(w) = F_{z,w}^\lambda \left(a(z)b(w)\right) = \sum_{n=0}^\infty \lambda^{(n)} \left(a(w)_n b(w)\right).$$

As before, we have the following formula for n-*th product*:

$$(2.3.10b) \qquad a(w)_n b(w) = \mathrm{Res}_z (z - w)^n a(z)b(w).$$

In the case when U is a Lie (super)algebra we will use the bracket notation for the λ-product, will call it the λ-bracket and will denote by $a(w)_{(n)} b(w)$ the n-th product (given by (2.3.8)), i.e.:

$$(2.3.11) \qquad [a(w)_\lambda b(w)] = \sum_{m=0}^\infty \lambda^{(m)} \left(a(w)_{(m)} b(w)\right).$$

The following formulas are very useful in studying associativity properties of the λ-product:

$$(2.3.12) \qquad F_{z,w}^\lambda F_{x,w}^\mu a(z)(b(x)c(w)) = a(w)_\lambda \left(b(w)_\mu c(w)\right),$$

$$(2.3.13) \qquad F_{z,w}^\lambda F_{x,w}^\mu (a(z)b(x))c(w) = (a(w)_\lambda b(w))_{\lambda+\mu} c(w).$$

The first of these two formulas is obvious, while the second is immediate by (2.2.12).

Now we can prove the basic properties of λ-products and λ-brackets.

PROPOSITION 2.3. (a) *For any two U-valued formal distributions $a(w)$ and $b(w)$, where U is an arbitrary algebra, one has:*

$$(\partial_w a(w))_\lambda \, b(w) = -\lambda a(w)_\lambda b(w),$$

$$a(w)_\lambda \partial_w b(w) = (\lambda + \partial_w)\left(a(w)_\lambda b(w)\right).$$

In particular, ∂_w is a derivation of the λ-product.

(b) *Let $a(w)$ and $b(w)$ be U-valued formal distributions, where U is an arbitrary algebra, such that the formal distribution $a(z)b(w)$ is local. Let $a(w) \underset{\lambda}{\circ} b(w)$ denote the λ-product of the algebra U^{op} (which is U with the opposite multiplication $a \circ b = ba$). Then*

$$b(w) \underset{\lambda}{\circ} a(w) = a(w)_{-\lambda-\partial_w} b(w).$$

(c) *Let $a(w)$ and $b(w)$ be $\mathfrak{g}$-valued mutually local formal distributions, where $\mathfrak{g}$ is a Lie superalgebra (i.e. (2.3.2) holds). Then*

$$[a(w)_\lambda b(w)] = -p(a,b)\,[b(w)_{-\lambda-\partial_w}\, a(w)].$$

(d) *Let $a(w)$, $b(w)$ and $c(w)$ be $\mathfrak{g}$-valued formal distributions, where $\mathfrak{g}$ is a Lie superalgebra. Then*

$$[a(w)_\lambda\, [b(w)_\mu c(w)]] = [[a(w)_\lambda b(w)]_{\lambda+\mu}\, c(w)] + p(a,b)\,[b(w)_\mu\, [a(w)_\lambda c(w)]].$$

PROOF. (a) follows from (2.2.10) applied to the formal distribution $a(z,w) = a(z)b(w)$. (b) follows similarly from (2.2.11). (c) is immediate by (b). Finally (d) follows from (2.3.12) and (2.3.13) applied to the Jacobi identity:

$$[a(z), [b(x), c(w)]] = [[a(z), b(x)]\,, c(w)] + p(a,b)\,[b(x), [a(z), c(w)]].$$

$\square$

REMARK 2.3b. (i) Proposition 2.3 (a) in terms of n-th products means the following formulas (cf. Remark 2.2a):

(2.3.14a) $$\partial a(w)_n b(w) = -n a(w)_{n-1} b(w),$$

(2.3.14b) $$a(w)_n \partial b(w) = \partial\,(a(w)_n b(w)) + n a(w)_{n-1} b(w).$$

Hence ∂_w is a derivation of all n-th products.

(ii) Proposition 2.3 (c) in terms of n-th products means (cf. Remark 2.2b):

(2.3.15) $$a(w)_{(n)} b(w) = -p(a,b) \sum_{j=0}^{\infty} (-1)^{j+n} \partial_w^{(j)}\,(b(w)_{(n+j)} a(w)),$$

provided that $a(w)$ and $b(w)$ are mutually local.

(iii) Proposition 2.3 (d) in terms of n-th products means

$$(2.3.16) \quad a(w)_{(m)}\big(b(w)_{(n)}c(w)\big) = \sum_{j=0}^{m} \binom{m}{j}\big(a(w)_{(j)}b(w)\big)_{(m+n-j)}c(w)$$
$$+\, p(a,b)b(w)_{(n)}\big(a(w)_{(m)}c(w)\big)\ .$$

The following well-known statement has many important applications.

COROLLARY 2.3. *Let $\mathfrak{g}$ be a Lie superalgebra.*

(a) *If $a(z)$ and $b(z)$ are $\mathfrak{g}$-valued formal distributions, then $\big[a_{(0)}, b(z)\big] = 0$ iff $a(z)_{(0)}b(z) = 0$.*

(b) *If $a(z)$ is an odd $\mathfrak{g}$-valued formal distribution, then $a_{(0)}^2 = 0$ iff $\mathrm{Res}_z\, a(z)_{(0)}a(z) = 0$.*

(c) *Let $\mathcal{A}$ be a space consisting of mutually local formal $\mathfrak{g}$-valued distributions in w which is ∂-invariant and closed with respect to all n-th products, $n \in \mathbb{Z}_+$. Then with respect to the 0-th product $\partial\mathcal{A}$ is a 2-sided ideal of $\mathcal{A}$ and $\mathcal{A}/\partial\mathcal{A}$ is a Lie superalgebra. Moreover, the 0-th product defines on $\mathcal{A}$ a structure of a left $\mathcal{A}/\partial\mathcal{A}$-module.*

PROOF. Statements (a) and (b) are obvious by definitions. From (2.3.14a) and (2.3.14b) for $n = 0$ we get

$$(2.3.17) \qquad\qquad (\partial\mathcal{A})_{(0)}\mathcal{A} = 0, \quad \mathcal{A}_{(0)}\partial\mathcal{A} \subset \mathcal{A},$$

Hence $\partial\mathcal{A}$ is a 2-sided ideal. Furthermore, (2.3.15) for $n = 0$ gives

$$(2.3.18) \qquad\qquad a(w)_{(0)}b(w) = -p(a,b)b(w)_{(0)}a(w) \bmod \partial\mathcal{A}.$$

Hence the 0-th product induces a super skew-symmetric bracket on $\mathcal{A}/\partial\mathcal{A}$. The super Jacobi identity in $\mathcal{A}/\partial\mathcal{A}$ follows from (2.3.16) for $m = n = 0$. This proves (c). $\qquad\square$

2.4. Taylor's formula

One of the devices in calculating the OPE is Taylor's formula. Here and further we shall adopt the following notational conventions. Given a formal distribution $a(z) = \sum_n a_n z^n$ we may construct a formal distribution in z and w:

$$i_{z,w}a(z-w) := \sum_n a_n i_{z,w}(z-w)^n.$$

In order to further simplify notation we shall often say instead that we consider the formal distribution $a(z - w)$ in z and w in the domain $|z| > |w|$.

PROPOSITION 2.4 (Taylor's formula). *Let $a(z)$ be a formal distribution. Then one has the following equality of formal distributions in z and in w in the domain $|z| > |w|$:*

$$(2.4.1) \qquad a(z + w) = \sum_{j=0}^{\infty} \partial^{(j)} a(z) w^j.$$

PROOF. Let $a(z) = \sum_n a_n z^n$, so that $\partial^{(j)} a(z) = \sum_n \binom{n}{j} a_n z^{n-j}$. Comparing coefficients of a_n in (2.4.1), we need to show that

$$(2.4.2) \qquad (z + w)^n = \sum_{j=0}^{\infty} z^{n-j} w^j \binom{n}{j}.$$

But (2.4.2) is the binomial expansion in the domain $|z| > |w|$. $\qquad\square$

Replacing z by w and w by $z - w$ in (2.4.1) we get another form of Taylor's formula as an equality of formal distributions in w and $z - w$ in the domain $|z - w| < |w|$:

$$(2.4.3) \qquad a(z) = \sum_{j=0}^{\infty} \partial^{(j)} a(w) (z - w)^j.$$

The following, yet another version of Taylor's formula, shows that when calculating the singular part of the OPE one can use Taylor's expansion up to the required order.

THEOREM 2.4. *Let $a(z)$ be a formal distribution and N be a non-negative integer. Then one has the following equality of formal distributions in z and w:*

$$(2.4.4) \qquad \partial_w^N \delta(z - w) a(z) = \partial_w^N \delta(z - w) \sum_{j=0}^{N} \partial^{(j)} a(w) (z - w)^j.$$

PROOF. It suffices to check that for an arbitrary Laurent polynomial $f(z)$ one has:

$$\mathrm{Res}_z\, \partial_z^N \delta(z - w) a(z) f(z)$$
$$= \sum_{j=0}^{N} \partial^{(j)} a(w)\, \mathrm{Res}_z\, \partial_z^N \delta(z - w) (z - w)^j f(z).$$

Integrating by parts N times transforms this to the equality

$$\mathrm{Res}_z\, \delta(z-w)\partial^N(a(z)f(z)) = \sum_{j=0}^{N} \partial^{(j)}a(w)\, \mathrm{Res}_z\, \delta(z-w)\partial_z^N((z-w)^j f(z))$$

which, due to (2.1.7) and Leibnitz rule, is

$$\partial^N(a(w)f(w)) = \sum_{j=0}^{N} \partial^j a(w)\binom{N}{j}\partial^{N-j}f(w).$$

This holds by Leibnitz rule. $\qquad\square$

2.5. Current algebras

Here we discuss one of the most important examples of algebras spanned by mutually local formal distributions—the current algebras.

First we consider the simplest case—the oscillator algebra $\mathfrak{s}$. This is a Lie algebra with basis α_n $(n \in \mathbb{Z})$, K and the following commutation relations:

$$(2.5.1) \qquad\qquad [\alpha_m, \alpha_n] = m\delta_{m,-n}K, \quad [K, \alpha_m] = 0.$$

Consider the following $\mathfrak{s}$-valued formal distribution:

$$\alpha(z) = \sum_{n\in\mathbb{Z}} \alpha_n z^{-n-1}.$$

Then it is straightforward to check that

$$(2.5.2) \qquad\qquad [\alpha(z), \alpha(w)] = \partial_w \delta(z-w)K$$

(this follows also from the equivalence of (i) and (iv) of Theorem 2.3). In other words, the formal distribution $\alpha(z)$ is local (with respect to itself) with the OPE, considered in the universal enveloping algebra of $\mathfrak{s}$:

$$(2.5.3) \qquad\qquad \alpha(z)\alpha(w) \sim \frac{K}{(z-w)^2}.$$

The (even) formal distribution $\alpha(z)$ is usually called a *free boson*.

The current algebra is a non-abelian generalization of the oscillator algebra. Let $\mathfrak{g}$ be a Lie superalgebra with an invariant supersymmetric bilinear form $(.|.)$. "Invariant" means

$$([a,b]\,|c) = (a|\,[b,c]), \quad a,b,c \in \mathfrak{g},$$

and "supersymmetric" means

$$(a|b) = (-1)^{p(a)}(b|a) \quad \text{(in particular, } (\mathfrak{g}_{\bar{0}}|\mathfrak{g}_{\bar{1}}) = 0) \ .$$

The *loop algebra* associated to $\mathfrak{g}$ is the Lie superalgebra

$$\tilde{\mathfrak{g}} = \mathfrak{g}\left[t, t^{-1}\right] \left(= \mathfrak{g} \otimes_{\mathbb{C}} \mathbb{C}\left[t, t^{-1}\right]\right)$$

over $\mathbb{C}$ with $\mathbb{Z}/2\mathbb{Z}$ grading extending that of $\mathfrak{g}$ by $p(t) = \bar{0}$, and commutation relations ($m, n \in \mathbb{Z}; a, b \in \mathfrak{g}$):

$$[a_m, b_n] = [a, b]_{m+n} \ .$$

Here and further a_m stands for $a \otimes t^m$. Note that $\tilde{\mathfrak{g}}$ is the Lie superalgebra of regular maps of $\mathbb{C}^{\times}$ to $\mathfrak{g}$ (hence the name "loop algebra").

The *affinization* of the pair $(\mathfrak{g}, (.|.))$ is a central extension of the loop algebra $\tilde{\mathfrak{g}}$ by a 1-dimensional even center $\mathbb{C}K$:

$$\widehat{\mathfrak{g}} = \tilde{\mathfrak{g}} + \mathbb{C}K$$

defined by the commutation relations ($m, n \in \mathbb{Z}; a, b \in \mathfrak{g}$):

$$(2.5.4) \qquad [a_m, b_n] = [a, b]_{m+n} + m(a|b)\delta_{m,-n}K, \ [K, \widehat{\mathfrak{g}}] = 0.$$

The Lie superalgebra $\widehat{\mathfrak{g}}$ is usually called by physicists a *current algebra*. Note that loop algebra is a special case of a current algebra when the bilinear form $(.|.)$ is zero. If $\mathfrak{g}$ is a simple finite-dimensional Lie algebra with the (normalized) Killing form $(.|.)$, then $\widehat{\mathfrak{g}}$ is known as the affine Kac-Moody algebra [**K2**]. If $\mathfrak{g}$ is the 1-dimensional Lie algebra with a non-degenerate bilinear form, then we recover the example of the oscillator algebra.

Introduce the following formal distributions with values in $\widehat{\mathfrak{g}}$ which are usually called *currents*:

$$a(z) = \sum_{n \in \mathbb{Z}} a_n z^{-n-1}, \quad a \in \mathfrak{g}.$$

Then by the equivalence of (i) and (iv) of Theorem 2.3, we see that

$$(2.5.5) \qquad [a(z), b(w)] = \delta(z - w)[a, b](w) + \partial_w \delta(z - w)(a|b)K,$$

hence all the currents $a(z)$ are mutually local with the OPE, considered in the universal enveloping algebra $U(\widehat{\mathfrak{g}})$ of $\widehat{\mathfrak{g}}$:

$$(2.5.6) \qquad a(z)b(w) \sim \frac{[a,b]\,(w)}{z-w} + \frac{(a|b)K}{(z-w)^2}.$$

There exists a natural super extension of the affinization, called the *superaffinization*, which is a central extension of the super loop algebra (called a *supercurrent algebra*):

$$\widehat{\mathfrak{g}}_{\text{super}} = \mathfrak{g} \otimes_{\mathbb{C}} \mathbb{C}\left[t, t^{-1}, \theta\right] + \mathbb{C}K,$$

where $\theta^2 = 0$, $p(\theta) = \bar{1}$ and the remaining OPE are as follows. For $a \in \mathfrak{g}$ define the *supercurrent*

$$\bar{a}(z) = \sum_{n\in\mathbb{Z}} a_{n+\frac{1}{2}} z^{-n-1},$$

where $a_{n+\frac{1}{2}} = a \otimes t^n \theta$. Then the supercurrents $\bar{a}(z)$ are mutually local and also local with respect to the currents, and the remaining OPE are given by

$$(2.5.7a) \qquad a(z)\bar{b}(w) \quad \sim \quad \frac{\overline{[a,b]}(w)}{z-w},$$

$$(2.5.7b) \qquad \bar{a}(z)\bar{b}(w) \quad \sim \quad \frac{(b|a)K}{z-w}.$$

The supercurrents form a closed (under OPE) subalgebra. In view of its importance, we repeat its construction in a slightly different form. Let A be a superspace with a skew-supersymmetric bilinear form, i.e.,

$$(\varphi|\psi) = -(-1)^{p(\varphi)}(\psi|\varphi) \quad \text{(in particular, } (A_{\bar{0}}|A_{\bar{1}}) = 0).$$

The *Clifford affinization* of $(A, (.|.))$ is the Lie superalgebra

$$C_A = A\left[t, t^{-1}\right] + \mathbb{C}K$$

with commutation relations $(m, n \in \frac{1}{2} + \mathbb{Z}; \varphi, \psi \in A)$

$$(2.5.8) \qquad [\varphi_m, \psi_n] = (\varphi|\psi)\delta_{m,-n}K, \quad [C_A, K] = 0,$$

where $\varphi_m = \varphi \otimes t^{m-\frac{1}{2}}$. The formal distributions $\varphi(z) = \sum_{n\in\mathbb{Z}} \varphi_{n+\frac{1}{2}} z^{-n-1}$ are mutually local with the OPE (in the universal enveloping algebra of C_A):

$$(2.5.9) \qquad \varphi(z)\psi(w) \sim \frac{(\varphi|\psi)K}{z-w}.$$

Two particularly important special cases of the Clifford affinization are the following.

Let A be the odd 1-dimensional superspace $\mathbb{C}\varphi$ with the bilinear form $(\varphi|\varphi) = 1$, and let $K = 1$. Then C_A turns into the algebra

$$(2.5.10) \qquad \varphi_m\varphi_n + \varphi_n\varphi_m = \delta_{m,-n}, \quad m, n \in \frac{1}{2} + \mathbb{Z}.$$

The (odd) formal distribution $\varphi(z) = \sum_{n\in\frac{1}{2}+\mathbb{Z}} \varphi_n z^{-n-1/2}$ is called a *neutral free fermion*; its OPE is

$$(2.5.11) \qquad \varphi(z)\varphi(w) \sim \frac{1}{z-w}.$$

In the second example let A be the odd 2-dimensional superspace $\mathbb{C}\varphi^+ \oplus \mathbb{C}\varphi^-$ with the symmetric bilinear form $(\varphi^+|\varphi^-) = 1$, $(\varphi^\pm|\varphi^\pm) = 0$, and again let $K = 1$. Then we obtain the algebra ($m, n \in \frac{1}{2} + \mathbb{Z}$):

$$(2.5.12) \qquad \varphi_m^\pm\varphi_n^\mp + \varphi_n^\mp\varphi_m^\pm = \delta_{m,-n}, \quad \varphi_m^\pm\varphi_n^\pm + \varphi_n^\pm\varphi_m^\pm = 0.$$

The odd formal distributions $\varphi^\pm(z) = \sum_{n\in\frac{1}{2}+\mathbb{Z}} \varphi_n^\pm z^{-n-1/2}$ are called *charged free fermions*; their OPE are:

$$(2.5.13) \qquad \varphi^\pm(z)\varphi^\mp(w) \sim \frac{1}{z-w}, \quad \varphi^\pm(z)\varphi^\pm(w) \sim 0.$$

These examples show that superalgebra is far from being a senseless generalization of the usual algebra.

2.6. Conformal weight and the Virasoro algebra

Let H be a diagonalizable derivation of the associative algebra U, called a *Hamiltonian*. Then H acts on the space of all formal distributions with values in U in the obvious way (coefficient-wise). The following definition is motivated by (1.2.6b).

DEFINITION 2.6a. A formal U-valued distribution $a = a(z, w, \dots)$ is called an *eigendistribution* for H of *conformal weight* $\Delta \in \mathbb{C}$ if

$$(H - \Delta - z\partial_z - w\partial_w - \cdots)a = 0.$$

Here are some obvious properties of conformal weights.

PROPOSITION 2.6. *Suppose a and b are eigendistributions of conformal weights Δ and Δ' respectively. Then*

(a) *$\partial_z a$ is an eigendistribution of conformal weight $\Delta + 1$.*

(b) *$: a(z)b(w) :$ is an eigendistribution of conformal weight $\Delta + \Delta'$.*

(c) *The n-th OPE coefficient of $[a(z), b(w)]$ is an eigendistribution of conformal weight $\Delta + \Delta' - n - 1 (n \in \mathbb{Z}_+)$.*

(d) *If f is a homogeneous function of degree j then fa is an eigendistribution of conformal weight $\Delta - j$.*

COROLLARY 2.6. *If $a(z)$ and $b(z)$ are mutually local eigendistributions of conformal weights Δ and Δ', then in the OPE*

$$a(z)b(w) \sim \sum_{j=0}^{N-1} \frac{c^j(w)}{(z-w)^{j+1}}$$

all the summands have the same conformal weight $\Delta + \Delta'$.

If $a(z)$ is an eigendistribution of conformal weight Δ, one usually writes it in the form (without parenthesis around indices):

$$a(z) = \sum_{n \in -\Delta + \mathbb{Z}} a_n z^{-n-\Delta}.$$

The condition of $a(z)$ being an eigendistribution of conformal weight Δ is then equivalent to

$$(2.6.1) \qquad\qquad Ha_n = -na_n.$$

As a result, the commutation relations given by Theorem 2.3(iv) take a graded form:

$$(2.6.2a) \qquad\qquad [a_m, b_n] = \sum_{j=0}^{N-1} \binom{m + \Delta - 1}{j} c^j_{m+n},$$

or equivalently

$$(2.6.2b) \qquad\qquad [a_m, b(z)] = \sum_{j=0}^{N-1} \binom{m + \Delta - 1}{j} c^j(z) z^{m+\Delta-j-1}.$$

EXAMPLE 2.6. Choosing for the algebra of currents $\widehat{\mathfrak{g}}$ (resp. supercurrents $\widehat{\mathfrak{g}}_{\text{super}}$) the Hamiltonian $H = -t\partial_t$ (resp. $= -t\partial_t - \frac{1}{2}\theta\partial_\theta$), we see that the currents $a(z)$ (resp. supercurrents $\bar{a}(z)$) have conformal weight 1 (resp. 1/2).

Corollary 2.6 is a very useful bookkeeping device in calculating the OPE. In many examples (e.g., from the considerations of unitarity) the conformal weight is in $\frac{1}{2}\mathbb{Z}_+$ and it is 0 iff the eigendistribution is a constant element commuting with all formal distributions of the theory.

If the above positivity condition holds, then due to Corollary 2.6, all mutually local eigendistributions of conformal weight $\frac{1}{2}$ have the OPE of the form (2.5.7b), all eigendistributions of conformal weight 1 have the OPE of the form (2.5.6) and the OPE between the latter and the former is given by (2.5.7a).

We consider now the next case—a local (i.e., local to itself) even eigendistribution $L(z)$ of conformal weight 2:

$$L(z) = \sum_{n\in\mathbb{Z}} L_n z^{-n-2}.$$

As has been mentioned above, it is natural to assume that the OPE has the form

$$(2.6.3) \qquad L(z)L(w) \sim \frac{\frac{1}{2}C}{(z-w)^4} + \frac{a(w)}{(z-w)^3} + \frac{2b(w)}{(z-w)^2} + \frac{c(w)}{z-w},$$

where C is a constant formal distribution.

THEOREM 2.6. *Suppose that $L(z)$ is an even local formal distribution with the OPE of the form (2.6.3). Then*

(a) $a(w) = 0$ and $c(w) = \partial b(w)$.

(b) *If in addition $[C, L(z)] = 0$ and*

$$(2.6.4) \qquad [L_{-1}, L(z)] = \partial L(z), \quad [L_0, L(z)] = (z\partial + 2)L(z)$$

then (2.6.3) becomes

$$(2.6.5) \qquad L(z)L(w) \sim \frac{\frac{1}{2}C}{(z-w)^4} + \frac{2L(w)}{(z-w)^2} + \frac{\partial L(w)}{z-w},$$

or, equivalently, we have the Virasoro algebra $(m, n \in \mathbb{Z})$:

$$(2.6.6) \qquad [L_m, L_n] = (m-n)L_{m+n} + \frac{m^3-m}{12}\delta_{m,-n}C, \quad [C, L_m] = 0.$$

PROOF. Exchanging z and w in (2.6.3), we obtain

$$L(w)L(z) \sim \frac{\frac{1}{2}C}{(z-w)^4} - \frac{a(z)}{(z-w)^3} + \frac{2b(z)}{(z-w)^2} - \frac{c(z)}{z-w}.$$

By making use of Taylor's formula, this turns into:

$$(2.6.7) \qquad L(w)L(z) \sim \frac{\frac{1}{2}C}{(z-w)^4} - \frac{a(w) + \partial a(w)(z-w) + \partial^{(2)}a(w)(z-w)^2}{(z-w)^3} + \frac{2b(w) + 2\partial b(w)(z-w)}{(z-w)^2} - \frac{c(w)}{z-w}.$$

Due to locality the right-hand sides of (2.6.3) and (2.6.7) must be equal. Matching the coefficients of $(z-w)^{-3}$ and $(z-w)^{-1}$ we get (a). Thus, we have:

$$(2.6.8) \qquad L(z)L(w) \sim \frac{\frac{1}{2}C}{(z-w)^4} + \frac{2b(w)}{(z-w)^2} + \frac{\partial b(w)}{z-w}.$$

Due to (2.6.2b) this implies, in particular:

$$[L_{-1}, L(z)] = \partial b(z), \quad [L_0, L(z)] = (z\partial + 2)b(z).$$

Hence assumptions (2.6.4) imply that $b(z) = L(z)$. This proves (2.6.5). The equation (2.6.6) is equivalent to this OPE due to (2.6.2a). $\qquad\square$

A local formal distribution $L(z)$ with the OPE (2.6.5) is called a *Virasoro formal distribution* with *central charge* C.

In Table OPE we give a table of the most commonly used OPE of mutually local formal distributions and the equivalent commutation relations (all these are special cases of formula (2.6.2a)).

The definition given below singles out the most important for conformal field theory Lie superalgebras, which includes the (super)current algebra and the Virasoro algebra.

DEFINITION 2.6b. A Lie superalgebra $\mathfrak{g}$ is called a *formal distribution Lie superalgebra* if it is spanned over $\mathbb{C}$ by coefficients of a family F of $\mathfrak{g}$-valued mutually local formal distributions.

For example, the Virasoro algebra with $F = \{L(z), C\}$ and the current algebra $\widehat{\mathfrak{g}}$ with $F = \{a(z) \text{ where } a \in \mathfrak{g}, K\}$ are formal distribution Lie (super)algebras. We shall often write $(\mathfrak{g}, F)$ in order to emphasize the dependence on F.

Note that formal distribution Lie superalgebras form a category with morphisms $(\mathfrak{g}, F) \to (\mathfrak{g}_1, F_1)$ being homomorphisms $\varphi : \mathfrak{g} \to \mathfrak{g}_1$ such $\varphi(F) \subset \overline{F}_1$, where $\overline{F}_1$ is the closure of F_1, defined in the next section.

Table OPE.

1st distribution	2nd distribution	commutation relations	OPE
$a(z) = \sum a_m z^{-m-1}$	$b(w) = \sum b_n w^{-n-1}$	$[a_m, b_n] = c_{m+n}$	$\dfrac{c(w) = \sum c_n w^{-n-1}}{z - w}$
$a(z) = \sum a_m z^{-m-1}$	$b(w) = \sum b_n w^{-n-1}$	$[a_m, b_n] = m\delta_{m,-n}$	$\dfrac{1}{(z - w)^2}$
$L(z) = \sum L_m z^{-m-2}$	$a(w) = \sum a_n w^{-n-\Delta}$	$[L_m, a_n] = ((\Delta - 1)m - n)a_{m+n}$	$\dfrac{\partial a(w)}{z - w} + \dfrac{\Delta a(w)}{(z - w)^2}$
$L(z) = \sum L_m z^{-m-2}$	$L(w) = \sum L_n w^{-n-2}$	$[L_m, L_n] = (m - n)L_{m+n}$ $+ \dfrac{m^3 - m}{12}\delta_{m,-n}c$	$\dfrac{\partial L(w)}{z - w} + \dfrac{2L(w)}{(z - w)^2}$ $+ \dfrac{c/2}{(z - w)^4}$

2.7. Formal distribution Lie superalgebras and conformal superalgebras

This and the next two sections is an introduction to the theory of conformal (super)algebras. Though they are ideologically closely related to the theory of vertex algebras, the rest of the book may be read independently of them, except for the last Section 5.10.

Let $\mathfrak{g}$ be an arbitrary Lie superalgebra. We denote by $fd(\mathfrak{g})$ the space of all $\mathfrak{g}$-valued formal distributions in z endowed with n-th products (2.3.8), $n \in \mathbb{Z}_+$. This is also a $\mathbb{C}[\partial]$-module ($\partial = \partial_z$).

Consider the subspace R over $\mathbb{C}$ of $fd(\mathfrak{g})$ which is closed under all n-th products, $n \in \mathbb{Z}_+$, and denote by $\mathfrak{g}(R)$ the $\mathbb{C}$-space of all coefficients of all formal distributions from R. Provided that all formal distributions from R are mutually local, $\mathfrak{g}(R)$ is a subalgebra of $\mathfrak{g}$ with the bracket

$$(2.7.1) \qquad [a_{(m)}, b_{(n)}] = \sum_{j \in \mathbb{Z}_+} \binom{m}{j} (a_{(j)}b)_{(m+n-j)}.$$

This follows from Theorem 2.3(iv). Clearly, $\mathfrak{g}(R)$ is a formal distribution Lie superalgebra and all of them are thus obtained.

Let F be a collection of mutually local formal distributions from $fd(\mathfrak{g})$. We denote by $\bar{F}$ the *closure* of F, defined as the minimal $\mathbb{C}[\partial]$-submodule of $fd(\mathfrak{g})$ closed under all n-th products, $n \in \mathbb{Z}_+$. Due to Lemma 2.8 proved in Section 2.8 (applied to the adjoint representation) and Remark 2.3a, $\bar{F}$ consists of mutually local formal distributions and therefore we have a formal distribution Lie superalgebra $\mathfrak{g}(\bar{F})$. In view of Proposition 2.3 (or rather Remark 2.3b), this leads us to the following definition.

DEFINITION 2.7. A *conformal superalgebra* R is a left $\mathbb{Z}/2\mathbb{Z}$-graded $\mathbb{C}[\partial]$-module $R = R_{\bar{0}} \oplus R_{\bar{1}}$ with a $\mathbb{C}$-bilinear product $a_{(n)}b$ for each $n \in \mathbb{Z}_+$ such that the following axioms hold ($a, b, c \in R$, $m, n \in \mathbb{Z}_+$):

(C0) $a_{(n)}b = 0$ for $n \gg 0$,

(C1) $(\partial a)_{(n)}b = -na_{(n-1)}b$,

(C2) $a_{(n)}b = -p(a,b) \sum_{j=0}^{\infty} (-1)^{j+n} \partial^{(j)} \left(b_{(n+j)}a \right)$,

$$\text{(C3)} \quad a_{(m)}\left(b_{(n)}c\right) = \sum_{j=0}^{m} \binom{m}{j} (a_{(j)}b)_{(m+n-j)}c + p(a,b)b_{(n)}(a_{(m)}c) .$$

Note that axioms (C1) and (C2) imply

$$\text{(C1')} \qquad\qquad a_{(n)}\partial b = \partial(a_{(n)}b) + na_{(n-1)}b,$$

hence ∂ is a derivation of all n-th products (cf. Proposition 2.3(a)).

REMARK 2.7a. The operator $a_{(0)}$ is a derivation of all n-th products (due to (C3)) and it commutes with ∂ (due to (C1')). As in the proof of Corollary 2.3, it follows (using also (C1) and (C2)) that, with respect to 0-th product, ∂R is a 2-sided ideal of R such that $R/\partial R$ is a Lie superalgebra, and that 0-th product defines on R a structure of a left $R/\partial R$-module for which $R/\partial R$ commutes with $\mathbb{C}[\partial]$.

The notions of a homomorphism, ideal and subalgebra of a conformal superalgebra R are defined in the usual way. Conformal superalgebras form a category with morphisms being homomorphisms. An element $a \in R$ is called *central* if $a_{(n)}R = 0$ for all $n \in \mathbb{Z}_+$ (and hence $R_{(n)}a = 0$, $n \in \mathbb{Z}_+$). A conformal superalgebra is called *finite* if it is finitely generated as a $\mathbb{C}[\partial]$-module.

An efficient way to handle the n-th products of a conformal superalgebra R is to introduce the λ-*bracket* (cf. Section 2.3):

$$\text{(2.7.2)} \qquad\qquad [a_\lambda b] = \sum_{n=0}^{\infty} \lambda^{(n)} \left(a_{(n)}b\right) .$$

Here λ is an indeterminate and, as before, $\lambda^{(n)}$ stands for $\lambda^n/n!$. Due to axiom (C0), the λ-bracket defines a $\mathbb{C}$-linear map

$$R \otimes_{\mathbb{C}} R \to \mathbb{C}[\lambda] \otimes_{\mathbb{C}} R.$$

Due to Proposition 2.3 and Remark 2.3a axioms (C1)-(C3) are equivalent respectively to

$$\text{(C1)}_\lambda \qquad\qquad [\partial a_\lambda b] = -\lambda\,[a_\lambda b],$$

$$\text{(C2)}_\lambda \qquad\qquad [a_\lambda b] = -p(a,b)\,[b_{-\lambda-\partial}a],$$

$$\text{(C3)}_\lambda \qquad\qquad [a_\lambda[b_\mu c]] - p(a,b)\,[b_\mu[a_\lambda c]] = [[a_\lambda b]_{\lambda+\mu}c].$$

Axioms $(C1)_\lambda$ and $(C2)_\lambda$ imply

$$(C1')_\lambda \qquad\qquad [a_\lambda \partial b] = (\partial + \lambda)\,[a_\lambda b]\,,$$

hence ∂ is a derivation of the λ-bracket.

The first application of the λ-product is the following corollary.

COROLLARY 2.7. [**DK**] *Any torsion element a of a finite conformal superalgebra R is central. In particular, if R is finite, then, as a $\mathbb{C}[\partial]$-module, R is a direct sum of a finite-dimensional (over $\mathbb{C}$) central subalgebra and a free $\mathbb{C}[\partial]$-module of finite rank.*

PROOF. By definition, we have $P(\partial)a = 0$ for some polynomial P, hence $[P(\partial)a_\lambda b] = 0$ for any $b \in R$, and $P(-\lambda)\,[a_\lambda b] = 0$ by $(C1)_\lambda$. It follows that $[a_\lambda b] = 0$ for any $b \in R$, hence a is a central element. $\qquad\qquad\square$

Conformal superalgebras are an effective tool to study formal distribution Lie superalgebras. Indeed, if $\mathfrak{g}$ is spanned by coefficients of a collection F of mutually local formal distributions, then $\bar{F}$ is a conformal superalgebra, due to Proposition 2.3. Conversely, we may construct a formal distribution Lie superalgebra Lie R associated with a conformal superalgebra R as follows. Let Lie R be the quotient of the vector space with basis a_n ($a \in R$, $n \in \mathbb{Z}$) by the subspace spanned over $\mathbb{C}$ by elements:

$$(\lambda a)_n - \lambda a_n,\ \ (a+b)_n - a_n - b_n,\ \ (\partial a)_n + n a_{n-1},\ \ \text{where } a, b \in R, \lambda \in \mathbb{C}, n \in \mathbb{Z}.$$

One can check that a formula similar to (2.7.1) gives a well-defined bracket on Lie R:

$$(2.7.3) \qquad\qquad [a_m, b_n] = \sum_{j \in \mathbb{Z}_+} \binom{m}{j} \left(a_{(j)}b\right)_{m+n-j}.$$

Instead of doing this calculation, we shall use a more conceptual approach.

The *affinization* of a conformal superalgebra R is the conformal superalgebra

$$\tilde{R} = R\left[t, t^{-1}\right], p(t) = \bar{0},$$

with $\tilde{\partial} = \partial \otimes 1 + 1 \otimes \partial_t$ and the n-th product defined by ($a, b \in R$, $f, g \in \mathbb{C}\left[t, t^{-1}\right]$, $n \in \mathbb{Z}_+$):

$$(2.7.4) \qquad (a \otimes f)_{(n)}(b \otimes g) = \sum_{j \in \mathbb{Z}_+} (a_{(n+j)}b) \otimes ((\partial_t^{(j)} f)g).$$

(We shall see that the affinization of a conformal superalgebra is a straightforward generalization of a more naturally looking notion of affinization of a vertex algebra introduced by Borcherds; see Section 4.3.) Letting $a_n = a \otimes t^n$, formula (2.7.4) becomes ($m, n \in \mathbb{Z}$):

$$(2.7.5) \qquad (a_m)_{(k)}(b_n) = \sum_{j \in \mathbb{Z}_+} \binom{m}{j} \left(a_{(k+j)}b\right)_{m+n-j}.$$

Letting

$$\mathrm{Lie}\, R = \tilde{R}/\tilde{\partial}\tilde{R}$$

with the bracket induced by the 0-th product on $\tilde{R}$, (and keeping the notation a_n for its image in $\mathrm{Lie}\, R$) we obtain, due to Remark 2.7a, a well-defined Lie superalgebra, which is obviously the same as the one introduced above.

It remains to check the axioms of conformal superalgebra for $\tilde{R}$. A simple calculation shows that the corresponding λ-bracket is given by

$$(2.7.6) \qquad [a \otimes f_\lambda b \otimes g] = [a_{\lambda+\partial_t} b] \otimes f(t)g(t')|_{t'=t}.$$

The verification of axioms is now straightforward. Let us check, for example, axiom $(C2)_\lambda$:

$$
\begin{aligned}
[a \otimes f_\lambda b \otimes g] &= [a_{\lambda+\partial_t} b] \otimes f(t)g(t')|_{t'=t} \\
&= -p(a,b)\,[b_{-\lambda-\partial_t-\partial}a] \otimes f(t)g(t')|_{t'=t} \\
&= -p(a,b)\,\left[b_{-\lambda-\partial-\partial_t-\partial_{t'}+\partial_{t'}}a\right] \otimes g(t')f(t)\big|_{t=t'} \\
&= -p(a,b)\,\left[b \otimes g_{-\lambda-\tilde{\partial}}a \otimes f\right].
\end{aligned}
$$

REMARK 2.7b. It is clear from (2.7.5) that $-1 \otimes \partial_t$ is a derivation of the 0-th product of the conformal superalgebra $\tilde{R}$. Since this operator commutes with $\tilde{\partial}$, it induces a derivation T of the Lie superalgebra $\mathrm{Lie}\, R$, given by the formula:

$$T(a_n) = -na_{n-1}.$$

REMARK 2.7c. Let R be a conformal superalgebra and suppose that, as a $\mathbb{C}[\partial]$-module,

$$R \simeq (\mathbb{C}[\partial] \otimes V) \oplus C,$$

where V is a vector space (over $\mathbb{C}$) and C consists of torsion elements. Then the vector space $V\left[t, t^{-1}\right] \oplus C\, t^{-1}$ is complementary in $\tilde{R}$ to $\tilde{\partial}\tilde{R}$. Hence, as a vector space (over $\mathbb{C}$), $\operatorname{Lie} R \simeq V\left[t, t^{-1}\right] \oplus C$, where C is a central subalgebra of $\operatorname{Lie} R$ (by Corollary 2.7). It suffices to check this in two cases: 1) $\dim_{\mathbb{C}} V = 1$ and $C = 0$, 2) $V = 0$ and $\dim_{\mathbb{C}} C = 1$, when it is straightforward. It follows, in particular, that

$$T|_C = \partial,$$

where T is the derivation of $\operatorname{Lie} R$ defined by Remark 2.7b.

REMARK 2.7d. The construction of the Lie superalgebra $\operatorname{Lie} R$ can be generalized by taking an arbitrary commutative associative algebra A with a derivation δ and letting

$$\operatorname{Lie}_A R = (R \otimes A)/(\partial \otimes 1 + 1 \otimes \delta)(R \otimes A)$$

with the bracket

$$[a \otimes f,\, b \otimes g] = [a_{\delta_1} b] \otimes fg,$$

where $\delta_1(fg) = \delta(f)g$. The operator $-1 \otimes \delta$ on $R \otimes A$ induces a derivation of $\operatorname{Lie}_A R$, giving it a structure of a differential Lie superalgebra, cf. [**R1**].

Each element $a \in R$ gives rise to a formal distribution $a(z) = \sum_{n \in \mathbb{Z}} a_n z^{-n-1}$ with coefficients in $\operatorname{Lie} R$. We denote this family of formal distributions by $F(R)$. They obviously span $\operatorname{Lie} R$ and are mutually local since formula (2.7.5) for $k = 0$ is equivalent to

$$(2.7.7) \qquad [a(z), b(w)] = \sum_{j \in \mathbb{Z}_+} \left(a_{(j)}b\right)(w) \partial_w^{(j)} \delta(z - w)$$

and $a_{(j)}b = 0$ for $j \gg 0$. Hence $(\operatorname{Lie} R, F(R))$ is a formal distribution Lie superalgebra. We thus constructed a functor, from the category of conformal superalgebras to the category of formal distribution Lie superalgebras.

Note that $F(R)(\subset fd(\operatorname{Lie} R))$ is a conformal superalgebra and that the map $\varphi : R \to F(R)$ defined by $\varphi(a) = a(z)$ is a surjective homomorphism of conformal superalgebras. Indeed, φ preserves j-th products due to (2.7.7), and φ preserves the $\mathbb{C}[\partial]$-module structure since $(\partial a)_n = -na_{n-1}$, which means that $\partial_z a(z) = (\partial a)(z)$.

LEMMA 2.7. *If $a \in R$ and the element $a_{-1} \in \mathrm{Lie}\,R$ is 0, then $a = 0$. In particular, φ is an isomorphism of conformal superalgebras.*

PROOF. Define a linear map $\tilde{\psi} : \tilde{R} \to R$ of vector spaces over $\mathbb{C}$ by

$$\tilde{\psi}\left(at^{-j-1}\right) = \partial^{(j)}a, \ \tilde{\psi}\left(at^{j}\right) = 0, \quad \text{where } j \in \mathbb{Z}_+.$$

Then $\tilde{\psi}\left(\tilde{\partial}\tilde{R}\right) = 0$, hence $\tilde{\psi}$ induces a map $\psi : \mathrm{Lie}\,R \to R$ such that $\psi(a_{-1}) = a$. $\square$

Recall that to a formal distribution Lie superalgebra $(\mathfrak{g}, F)$ one canonically associates a conformal superalgebra $\mathrm{Conf}(\mathfrak{g}, F) = \bar{F}$. This gives us a functor from the category of formal distribution Lie superalgebras to the category of conformal superalgebras, which we denote by Conf. We also have constructed a functor in the opposite direction that canonically associates to a conformal superalgebra R a formal distribution Lie superalgebra $(\mathrm{Lie}\,R, F(R))$; we denote this functor by Lie. Due to Lemma 2.7, we have:

$$\mathrm{Conf}(\mathrm{Lie}\,R) \simeq R.$$

Furthermore, we have:

$$\mathrm{Lie}(\mathrm{Conf}(\mathfrak{g}, F)) = \left(\mathrm{Lie}\,\bar{F}, \bar{F}\right).$$

By the very definition, the Lie superalgebra $\mathfrak{g}$ is a quotient of $\mathrm{Lie}\,\bar{F}$ by an ideal that does not contain all the coefficients of a non-zero formal distribution from $\bar{F}$. Such an ideal is called an *irregular ideal* of $\mathrm{Lie}\,\bar{F}$. Conversely, if $\mathfrak{g}$ is obtained from $\mathrm{Lie}\,\bar{F}$ as a quotient by an irregular ideal, then $\mathrm{Conf}\,\mathfrak{g} \simeq \bar{F}$. The formal distribution Lie superalgebras $(\mathrm{Lie}\,\bar{F}, \bar{F})$ and $((\mathrm{Lie}\,\bar{F})/I, F)$ are called *equivalent*. Hence, it is natural to call $(\mathrm{Lie}\,R, F(R))$ the *maximal formal distribution Lie superalgebra* associated to the conformal superalgebra R.

So, the functor Conf induces a functor Conf' from the category of equivalence classes of formal distribution Lie superalgebras to the category of conformal superalgebras and the functor Lie induces a functor Lie' going in the opposite direction.

Thus we have proved the following result.

THEOREM 2.7. *The functor* Conf$'$ *and* Lie$'$ *are inverse of each other and establish equivalence between the category of equivalence classes of formal distribution Lie superalgebras and the category of conformal superalgebras*

A formal distribution Lie superalgebra $(\mathfrak{g}, F)$ is called *finite* if $\bar{F}$ is a finitely generated $\mathbb{C}[\partial]$-module. Theorem 2.7 reduces the classification of (finite) formal distribution Lie superalgebras to the classification of (finite) conformal superalgebras.

Due to Corollary 2.7, the description of finite conformal superalgebras splits into two problems:

1. describe conformal superalgebras that are free $\mathbb{C}[\partial]$-modules of finite rank;
2. find central extensions of conformal superalgebras from 1. with center being in torsion.

The first problem is reduced to solution of a finite system of functional equations on a finite set of polynomials in two indeterminates as follows.

Let $R = \bigoplus_{j=1}^{n} \mathbb{C}[\partial]\, a^j$ be a $\mathbb{Z}/2\mathbb{Z}$ graded $\mathbb{C}[\partial]$-module with $p(a^i)$, denoted by $p(i)$, and let $[a^i_\lambda a^j] = \sum_k Q^{ij}_k(\lambda, \partial)a^k$. These λ-brackets give rise to a structure of a conformal superalgebra on R if Q^{ij}_k $(i, j, k = 1, \ldots, n)$ are polynomials in λ and ∂ subject to the following relations that are equivalent to axioms $(C2)_\lambda$ and $(C3)_\lambda$ respectively:

$$(2.7.8) \qquad Q^{ij}_k(\lambda, \partial) = -(-1)^{p(i)p(j)}Q^{ji}_k(-\partial - \lambda, \partial),$$

$$(2.7.9) \quad \sum_{s=1}^{n}\left(Q^{jk}_s(\mu, \partial + \lambda)Q^{is}_t(\lambda, \partial) - (-1)^{p(i)p(j)}Q^{ik}_s(\lambda, \partial + \mu)Q^{j,s}_t(\mu, \partial)\right)$$
$$= \sum_{s=1}^{n} Q^{ij}_s(\lambda, -\lambda - \mu)Q^{sk}_t(\lambda + \mu, \partial).$$

Due to equivalence of $(C3)_\lambda$ to the Jacobi identity in Lie R, it suffices to check (2.7.9) for all triples $1 \le i \le j \le k \le n$ and $1 \le t \le n$.

It is clearly impossible to solve these equations directly for $n \ge 2$. Below a solution is presented for $n = 1$ and $R = R_{\bar{0}}$ (obtained jointly with Minoru Wakimoto).

We have: $R = \mathbb{C}[\partial]\,a$ and $[a_\lambda a] = Q(\lambda, \partial)a$, where $Q(\lambda, \partial)$ is a polynomial in λ and ∂ satisfying two equations:

$$(2.7.10) \qquad\qquad Q(\lambda, \partial) = -Q(-\partial - \lambda, \partial),$$

$$(2.7.11) \quad Q(\mu, \partial + \lambda)Q(\lambda, \partial) - Q(\lambda, \partial + \mu)Q(\mu, \partial) = Q(\lambda, -\lambda - \mu)Q(\lambda + \mu, \partial).$$

Let $Q(\lambda, \partial) = \sum_{j=0}^{r} c_j(\lambda)\partial^j$ with $c_r(\lambda) \neq 0$. Comparing coefficients of ∂^{2r-1} in (2.7.11) we obtain: $r(\lambda - \mu)c_r(\lambda)c_r(\mu) = 0$ if $r > 1$, a contradiction. Hence $Q(\lambda, \partial) = a(\lambda)\partial + b(\lambda)$. Letting $\lambda = \mu$ in (2.7.11), we get $Q(\lambda, -2\lambda)Q(2\lambda, \partial) = 0$, hence $Q(\lambda, -2\lambda) = 0$, which means that $b(\lambda) = 2\lambda a(\lambda)$, hence $Q(\lambda, \partial) = a(\lambda)(\partial + 2\lambda)$. Plugging this in (2.7.10), we see that $a(\lambda)$ is a constant. Since a can be chosen up to a non-zero constant factor, we arrive at two solutions: $Q(\lambda, \partial) = 0$ and $Q(\lambda, \partial) = \partial + 2\lambda$. In the first case we get a commutative conformal algebra (i.e. all products are trivial), and in the second case we arrive at the Virasoro conformal algebra discussed below.

Now we discuss briefly the second problem, the construction of central extensions: $\widehat{R} = R \oplus C$ where R and C are $\mathbb{C}[\partial]$-submodules of $\widehat{R}$ and $C_\lambda \widehat{R} = 0$. The λ-bracket $[a_\lambda b]^\wedge$ on $R \subset \widehat{R}$ is given by

$$(2.7.12) \qquad\qquad [a_\lambda b]^\wedge = [a_\lambda b] + \alpha_\lambda(a, b),$$

where $[a_\lambda b]$ is the λ-product on R and $\alpha_\lambda(a, b) = \sum_{n \geq 0} \lambda^{(n)}\alpha_n(a, b)$ is a $\mathbb{C}$-linear map $R \otimes R \to \mathbb{C}[\lambda] \otimes_\mathbb{C} C$. The axioms $(C1)_\lambda$, $(C1')_\lambda$, $(C2)_\lambda$ and $(C3)_\lambda$ for $\widehat{R}$ are equivalent to the following properties of the 2-*cocycle* $\alpha_\lambda(a, b)$:

$$(2.7.13) \qquad \alpha_\lambda(\partial a, b) = -\lambda\alpha_\lambda(a, b), \quad \alpha_\lambda(a, \partial b) = (\partial + \lambda)\alpha_\lambda(a, b),$$

$$(2.7.14) \qquad\qquad \alpha_\lambda(a, b) = -p(a, b)\alpha_{-\lambda - \partial}(b, a),$$

$$(2.7.15) \qquad \alpha_\lambda(a, b_\mu c) - p(a, b)\alpha_\mu(b, a_\lambda c) = \alpha_{\lambda + \mu}(a_\lambda b, c).$$

As above, these equations are equivalent to a system of functional equations on a set of polynomials in two indeterminates. If we take another complement of C in $\widehat{R}$ by replacing $a \in R$ by $a - f(a)$, where $f : R \to C$ is a $\mathbb{C}[\partial]$-module homomorphism, then $\alpha_\lambda(a, b)$ gets replaced by $\alpha'_\lambda(a, b) = \alpha_\lambda(a, b) + f(a_\lambda b)$. The *trivial 2-cocycle*

$f(a_\lambda b)$ defines a trivial extension, and *equivalent 2-cocycles* $\alpha'_\lambda(a,b)$ and $\alpha_\lambda(a,b)$ define isomorphic extensions.

One can develop a cohomology theory of conformal superalgebras similar to the Lie algebra cohomology (see Section 2.11). The central extensions of R by C are then parameterized by $H^2(R,C)$.

We consider now three main examples of finite conformal (super)algebras R. Due to (C1) and (C1$'$) it suffices to define n-th products on the generators of the $\mathbb{C}[\partial]$-module R.

EXAMPLE 2.7a. Let $\mathfrak{g}$ be a finite-dimensional Lie superalgebra. Recall (see Section 2.5) that the associated loop algebra $\tilde{\mathfrak{g}} = \mathfrak{g}\left[t, t^{-1}\right]$ is a formal distribution Lie superalgebra with the family F consisting of currents $a(z) = \sum_n (at^n) z^{-n-1}$ where $a \in \mathfrak{g}$. Recall that (cf. (2.5.5)):

$$[a(z), b(w)] = [a, b]\,(w)\delta(z - w).$$

Hence the conformal superalgebra associated to $(\tilde{\mathfrak{g}}, F)$ is $\mathbb{C}[\partial] \otimes_\mathbb{C} \mathfrak{g}$ with a structure of a conformal superalgebra defined on $a, b \in \mathfrak{g}$ by

$$(2.7.16) \qquad a_{(0)}b = [a, b]\,, \quad a_{(m)}b = 0 \text{ for } m \geq 1.$$

This is called the *current conformal superalgebra* associated to $\mathfrak{g}$. It is denoted by Cur $\mathfrak{g}$.

The following formula defines a 2-cocycle on Cur $\mathfrak{g}$ with values in the trivial $\mathbb{C}[\partial]$-module $\mathbb{C}$ $(a, b \in 1 \otimes \mathfrak{g} \subset \text{Cur}\,\mathfrak{g})$:

$$(2.7.17) \qquad \alpha_1(a, b) = (a|b), \quad \alpha_m(a, b) = 0 \text{ if } m \neq 1,$$

where $(.|.)$ is a supersymmetric invariant bilinear form on $\mathfrak{g}$. It is easy to see that (2.7.17) gives all 2-cocycles, up to taking for α_0 a 2-cocycle on $\mathfrak{g}$, provided that $[\mathfrak{g}, \mathfrak{g}] = \mathfrak{g}$. In particular, if $\mathfrak{g}$ is a simple finite-dimensional Lie algebra, then (2.7.17) gives all 2-cocycles, up to equivalence. The corresponding central extension is the conformal superalgebra Conf $\widehat{\mathfrak{g}}$ associated to the current algebra $\widehat{\mathfrak{g}}$ defined in Section 2.5. It follows from Remark 2.7c that $\text{Lie}(\text{Conf}\,\widehat{\mathfrak{g}}) = \widehat{\mathfrak{g}}$ and $\text{Lie}(\text{Cur}\,\mathfrak{g}) = \tilde{\mathfrak{g}}$, i.e. both $\tilde{\mathfrak{g}}$ and $\widehat{\mathfrak{g}}$ are maximal formal distribution Lie superalgebras. Note that $I = \mathfrak{g}\left[t, t^{-1}\right] P(t)$, where $P(t)$ is a non-invertible Laurent polynomial, is an irregular ideal of $\tilde{\mathfrak{g}}$, hence the formal distribution Lie algebras $\tilde{\mathfrak{g}}$ and $\tilde{\mathfrak{g}}/I$ are equivalent (i.e. give rise to the same conformal algebra).

EXAMPLE 2.7b. Let Vect $\mathbb{C}^\times$ denote the Lie algebra of regular vector fields on $\mathbb{C}^\times$. It has a basis $L_n = -t^{n+1}\partial_t$ $(n \in \mathbb{Z})$ with commutation relations

$$[L_m, L_n] = (m - n)L_{m+n}.$$

This is a formal distribution Lie algebra with the family F consisting of a single formal distribution

$$L(z) = \sum_n L_n z^{-n-2} = -\delta(z - t)\partial_t.$$

Either directly (cf. Theorem 2.6) or using (2.1.10) we obtain:

$$[L(z), L(w)] = \partial_w L(w)\delta(z - w) + 2L(w)\partial_w \delta(z - w).$$

Hence the conformal algebra associated to $(\text{Vect}\,\mathbb{C}^\times, \{L(z)\})$ is $\text{Conf}(\text{Vect}\,\mathbb{C}^\times) = \mathbb{C}[\partial]\,L$, with products:

$$(2.7.18) \qquad L_{(0)}L = \partial L, \quad L_{(1)}L = 2L, \quad L_{(m)}L = 0 \text{ if } m \geq 2.$$

This is called the *Virasoro conformal algebra* and is denoted by Vir. It has been already encountered above in terms of the λ-bracket: $[L_\lambda L] = (\partial + 2\lambda)L$.

One can show that this conformal algebra has a unique, up to equivalence, 2-cocycle, which is given by

$$(2.7.19) \qquad \alpha_3(L, L) = \frac{c}{2}, \quad \alpha_m(L, L) = 0 \text{ if } m \neq 3.$$

The corresponding central extension is the conformal algebra $\text{Conf}(\text{Vir})$ associated to the Virasoro algebra (see Section 2.6). Note that both Vect $\mathbb{C}^\times$ and Vir are maximal formal distribution Lie algebras. Both have no irregular ideals.

EXAMPLE 2.7c. The obvious semidirect sum $(\text{Vect}\,\mathbb{C}^\times) + \tilde{\mathfrak{g}}$ defined by $[f(t)\partial_t, a \otimes g(t)] = a \otimes f(t)\partial_t g(t)$ is a maximal formal distribution Lie algebra with no irregular ideals. One has:

$$[L(z), a(w)] = (\partial_w a(w))\,\delta(z - w) + a(w)\partial_w \delta(z - w).$$

Hence the associated to $(\mathrm{Vect}\,\mathbb{C}^{\times}) + \tilde{\mathfrak{g}}$ conformal algebra is the semidirect sum $\mathrm{Conf}(\mathrm{Vect}\,\mathbb{C}^{\times}) + \mathrm{Cur}\,\mathfrak{g}$, defined by $(a \in \mathfrak{g})$:

$$(2.7.20) \qquad L_{(0)}a = \partial a, \quad L_{(1)}a = a, \quad L_{(m)}a = 0 \quad \text{for } m > 1.$$

In conclusion of this section we state without proofs the results of [**DK**] on classification of finite conformal algebras.

A conformal (super)algebra is called simple if it is not commutative and it contains no nontrivial ideals. The paper [**DK**] contains a proof of Conjecture 2.7 stated in the first edition of this book:

Any simple finite conformal algebra is isomorphic either to a current conformal algebra $\mathrm{Cur}\,\mathfrak{g}$, *where* $\mathfrak{g}$ *is a simple finite-dimensional Lie algebra, or to the Virasoro conformal algebra.*

Of course, translating this into the language of formal distribution Lie algebras, we obtain the following result: Any finite formal distribution Lie algebra which is simple (i.e. any its non-trivial ideal is irregular) is isomorphic either to $(\mathrm{Vect}\,\mathbb{C}^{\times}, \{L(z)\})$ or to a quotient of $(\tilde{\mathfrak{g}}, \{a(z)|a \in \mathfrak{g}\})$ where $\mathfrak{g}$ is a simple finite-dimensional Lie algebra.

The $\mathbb{C}$-span of all elements of the form $a_{(m)}b$ of a conformal (super)algebra R, $m \in \mathbb{Z}_+$, is called the *derived algebra* of R and is denoted by R'. It is easy to see that R' is an ideal of R such that R/R' is commutative. We have the derived series $R \supset R' \supset R'' \supset \ldots$. A conformal (super)algebra is called *solvable* if the n-th member of this series is zero for $n \gg 0$. A conformal (super)algebra is called *semisimple* if it contains no non-zero solvable ideals. The second main result of the paper [**DK**] states that *any finite semisimple conformal algebra is a direct sum of conformal algebras of the following types:*

(i) *current conformal algebra* $\mathrm{Cur}\,\mathfrak{g}$, *where* $\mathfrak{g}$ *is a semisimple finite-dimensional Lie algebra,*

(ii) *Virasoro conformal algebra,*

(iii) *the semidirect sum of (i) and (ii).*

The proof of these results uses heavily Cartan's theory of filtered Lie algebras.

As we shall see in Sections 5.9 and 5.10, the list of simple finite conformal superalgebras is much richer than that of conformal algebras.

2.8. Conformal modules and modules over conformal superalgebras

Let $\mathfrak{g}$ be a Lie superalgebra and let V be a $\mathfrak{g}$-module. We say that formal distributions $a(z) \in \mathfrak{g}\left[\left[z, z^{-1}\right]\right]$ and $v(z) \in V\left[\left[z, z^{-1}\right]\right]$ form a *local pair* if the formal distribution $a(z)v(w) \in V\left[\left[z, z^{-1}, w, w^{-1}\right]\right]$ is local, i.e.

$$(2.8.1) \qquad\qquad (z-w)^N a(z)v(w) = 0 \quad \text{for} \quad N \gg 0.$$

It follows from Corollary 2.2 that (2.8.1) is equivalent to

$$(2.8.2) \qquad a(z)v(w) = \sum_{j=0}^{N-1} \left(a(w)_{(j)}v(w)\right) \partial_w^{(j)}\delta(z-w),$$

where $a(w)_{(j)}v(w) \in V\left[\left[w, w^{-1}\right]\right]$ is defined by

$$(2.8.3) \qquad\qquad a(w)_{(j)}v(w) = \mathrm{Res}_z (z-w)^j a(z)v(w).$$

DEFINITION 2.8a. Let $(\mathfrak{g}, F)$ be a formal distribution Lie superalgebra and let V be a $\mathfrak{g}$-module spanned over $\mathbb{C}$ by coefficients of a family E of formal distributions such that all pairs $(a(z), v(z))$, where $a(z) \in F$ and $v(z) \in E$, are local. Then (V, E) is called a *conformal module* over $(\mathfrak{g}, F)$.

The following is a representation-theoretic analogue of Dong's lemma proved in Section 3.2.

LEMMA 2.8. *Let V be a module over a Lie superalgebra $\mathfrak{g}$, let $a(z), b(z) \in \mathfrak{g}\left[\left[z, z^{-1}\right]\right]$ and $v(z) \in V\left[\left[z, z^{-1}\right]\right]$.*
(a) If $(a(z), v(z))$ is a local pair, then both pairs $(\partial a(z), v(z))$ and $(a(z), \partial v(z))$ are local.
(b) If all three pairs $(a(z), b(z))$, $(a(z), v(z))$ and $(b(z), v(z))$ are local, then the pairs $\left(a(z)_{(j)}b(z), v(z)\right)$ and $\left(a(z), b(z)_{(j)}v(z)\right)$ are local for each $j \in \mathbb{Z}_+$.

PROOF. (a) is clear. In order to prove the first part of (b) we may assume that all three pairs satisfy (2.3.2) and (2.8.1) respectively for some $N \in \mathbb{Z}_+$. Then we have:

$$(z - w)^{3N} \left(a(z)_{(j)}b(z)\right) v(w)$$

$$= (z - w)^N \operatorname{Res}_u \sum_{i=0}^{2N} \binom{2N}{i} (z - u)^i (u - w)^{2N-i} (u - z)^j [a(u), b(z)] v(w).$$

The summation over i in the right-hand side may be replaced by that from 0 to N since $a(u)$ and $b(z)$ are mutually local. Hence it can be written as follows:

$$(z - w)^N \operatorname{Res}_u (u - w)^N P(z, u, w)(u - z)^j (a(u)b(z)v(w) - b(z)a(u)v(w))$$

for some polynomial P. But this is zero since both pairs (b, v) and (a, v) are local, which proves that the pair $\left(a_{(j)}b, v\right)$ is local.

Next, using the first part of lemma, we may find N for which all pairs $\left(b_{(j)}a, v\right)$ and (a, v) satisfy (2.8.1). Then we have:

$$a(z) \left(b(w)_{(j)}v(w)\right) = \operatorname{Res}_u a(z)b(u)v(w)(u - w)^j$$

$$= -\operatorname{Res}_u([b(u), a(z)]v(w) - b(u)a(z)v(w))(u - w)^j$$

$$= -\operatorname{Res}_u \left(\sum_{i \geq 0} \left(b(z)_{(i)}a(z)\right) v(w)\partial_z^{(i)} \delta(u - z) - b(u)a(z)v(w) \right)(u - w)^j,$$

hence $(z - w)^N a(z) \left(b(w)_{(j)}v(w)\right) = 0$. $\qquad\square$

Lemma 2.8 shows that the family E of a conformal module (V, E) over $(\mathfrak{g}, F)$ can always be included in its *closure*, i.e. the minimal family $\bar{E}$ which is still local with respect to F and such that $\mathbb{C}[\partial]\bar{E} \subset \bar{E}$ and $a_{(j)}\bar{E} \subset \bar{E}$ for all $a \in F$ and $j \in \mathbb{Z}_+$. The same lemma shows that $\bar{E}$ is local with respect to $\bar{F}$. Thus, we obtain the following corollary.

COROLLARY 2.8. (a) *If a Lie superalgebra $\mathfrak{g}$ is generated (as an algebra) by coefficients of a family of mutually local formal distributions F, then $(\mathfrak{g}, \bar{F})$ is a formal distribution Lie superalgebra.*

(b) *If V is a module over a formal distribution Lie superalgebra $(\mathfrak{g}, F)$, generated (as a module) by coefficients of a family E of formal distributions such that all pairs $(a(z), v(z))$, where $a(z) \in F$, $v(z) \in E$, are local, then $(V, \bar{E})$ is a conformal module over $(\mathfrak{g}, \overline{F})$.*

Note that conformal modules over a formal distribution Lie superalgebra $(\mathfrak{g}, F)$ form a category with morphisms $\varphi : (V, E) \to (V_1, E_1)$ being $\mathfrak{g}$-module homomorphisms $\varphi : V \to V_1$ such that $\varphi(E) \subset \overline{E}_1$.

The same calculation as in the proof of Proposition 2.3 gives for all $a(w), b(w) \in \mathfrak{g}\left[\left[w, w^{-1}\right]\right]$ and $v(w) \in V\left[\left[w, w^{-1}\right]\right]$ the following relations:

$$(2.8.4) \qquad \partial a(w)_{(n)} v(w) = \left[\partial_w, a(w)_{(n)}\right] v(w) = -n a(w)_{(n-1)} v(w),$$

$$(2.8.5) \qquad \left[a(w)_{(m)}, b(w)_{(n)}\right] v(w) = \sum_{j=0}^{m} \binom{m}{j} \left(a(w)_{(j)} b(w)\right)_{(m+n-j)} v(w).$$

(here $[,]$ is the bracket of operators on $\bar{E}$.) It follows from $(2.8.5)$ by induction on m (or from $(2.8.9)$ below) that $a_{(j)} \bar{E} \in \bar{E}$ for all $a \in \bar{F}$ and $j \in \mathbb{Z}_+$.

Thus, any conformal module (V, E) over a formal distribution Lie superalgebra $(\mathfrak{g}, F)$ gives rise to a *conformal module* $M(V) = \bar{E}$ over the conformal superalgebra $\bar{F}$, defined as follows.

DEFINITION 2.8b. A *module M* over a conformal superalgebra R is a left $\mathbb{Z}/2\mathbb{Z}$-graded $\mathbb{C}[\partial]$-module with $\mathbb{C}$-linear maps $a \mapsto a^M_{(n)}$ of R to $\mathrm{End}_{\mathbb{C}} M$ for each $n \in \mathbb{Z}_+$ such that the following properties hold for $a, b \in R$, $v \in M$, $m, n \in \mathbb{Z}_+$:

$$(M1) \qquad (\partial a)^M_{(n)} v = \left[\partial^M, a^M_{(n)}\right] v = -n a^M_{(n-1)} v,$$

$$(M2) \qquad \left[a^M_{(m)}, b^M_{(n)}\right] v = \sum_{j=0}^{m} \binom{m}{j} \left(a_{(j)} b\right)^M_{(m+n-j)} v.$$

An R-module M is called *conformal* if it satisfies the property

$$(M0) \qquad a^M_{(n)} v = 0 \quad \text{for} \quad n \gg 0.$$

REMARK 2.8a. We have in an arbitrary module M over a conformal superalgebra R: $(\partial R)_{(0)} M = 0$, hence the map $R_{(0)} M \to M$ endows M with the structure of a module over the Lie superalgebra $R/\partial R$ with respect to the 0-th product (cf. Remark 2.7a) which commutes with ∂^M. Thus, we get the $R/\partial R$-module $M/\partial^M M$.

Using this remark, conversely, as in Section 2.7, we canonically associate to a conformal module M over a conformal superalgebra R a conformal module $V(M)$ over the formal distribution Lie superalgebra Lie R as follows. First, we construct the affinization module $\tilde{M} = M\left[t, t^{-1}\right]$ over the conformal superalgebra $\tilde{R}$ by letting $\tilde{\partial}^{\tilde{M}} = \partial^M \otimes 1 + 1 \otimes \partial_t$ and defining for $a \in R$, $v \in M$, $f, g \in \mathbb{C}\left[t, t^{-1}\right]$, $n \in \mathbb{Z}_+$:

$$(2.8.6) \qquad (a \otimes f)^M_{(n)}(v \otimes g) = \sum_{j \in \mathbb{Z}_+} \left(a^M_{(n+j)}v\right) \otimes \left(\left(\partial_t^{(j)} f\right) g\right).$$

Then we let $V(M) = \tilde{M}/\tilde{\partial}^{\tilde{M}} \tilde{M}$ with the action of Lie R induced by the 0-th action:

$$(a \otimes f)(v \otimes g) = (a \otimes f)^M_{(0)}(v \otimes g).$$

Letting, as before, $a_n = a \otimes t^n$ and $v_n = v \otimes t^n$, we obtain from (2.8.6) an explicit formula for the action of Lie R on $V(M)$ ($a \in R$, $v \in M$, $m, n \in \mathbb{Z}$):

$$(2.8.7) \qquad a_m v_n = \sum_{j \in \mathbb{Z}_+} \binom{m}{j} \left(a^M_{(j)}v\right)_{m+n-j}.$$

The (Lie R, R)-module $(V(M), E(M))$, where

$$E(M) = \left\{v(z) = \sum_n v_n z^{-n-1} \middle| v \in M\right\},$$

is conformal, $E(M)$ being canonically isomorphic to the R-module M.

Proofs of these facts are similar to those in Section 2.7. The calculations, as before, are greatly simplified by introduction of the λ-*action* ($a \in R$, $v \in M$):

$$a^M_\lambda v = \sum_{n=0}^\infty \lambda^{(n)} a^M_{(n)} v \in M[[\lambda]].$$

Then axioms (M1) and (M2) become respectively:

$$(\text{M1})_\lambda \qquad (\partial a)^M_\lambda v = \left[\partial^M, a^M_\lambda\right] = -\lambda a^M_\lambda v,$$

$$(\text{M2})_\lambda \qquad \left[a^M_\lambda, b^M_\mu\right] v = [a_\lambda b]^M_{\lambda+\mu} v$$

Axiom (M0) means that $a^M_\lambda v \in \mathbb{C}[\lambda] \otimes_\mathbb{C} M$.

REMARK 2.8b. Replacing μ by $\mu - \lambda$ in (M2)$_\lambda$, we invert (M2)$_\lambda$:

$$(2.8.8) \qquad [a_\lambda b]^M_\mu v = \left[a^M_\lambda, b^M_{\mu-\lambda}\right] v.$$

Equivalently:

$$(2.8.9) \qquad \left(a_{(m)}b\right)^M_{(n)} v = \sum_{j=0}^m (-1)^{m+j} \binom{m}{j} \left[a^M_{(j)}, b^M_{(m+n-j)}\right] v.$$

Let R be a conformal superalgebra. We have constructed a functor from the category of conformal (Lie R)-modules to the category of conformal R-modules by sending (V, E) to $M(V) = \bar{E}$, and a functor in the opposite direction by sending M to $(V(M), E(M))$. As in Section 2.7, it is easy to see that these functors induce equivalence between the category of equivalence classes of conformal (Lie R)-modules and the category of conformal R-modules.

The following proposition is proved in the same way as Corollary 2.7.

PROPOSITION 2.8. *Let M be a module over a conformal superalgebra R. Then*
(a) *Any torsion element of R acts trivially on M.*
(b) *Any torsion element $v \in M$ is an invariant of R, i.e. $R^M_{(n)}v = 0$ for all $n \in \mathbb{Z}_+$.*

An R-module M is called *finite* if it is finitely generated as a $\mathbb{C}[\partial]$-module.

An example of a conformal R-module, is, of course, the *adjoint module R* given by $a \mapsto a^R_{(n)} = a_{(n)}$. It is finite iff R is finite.

We consider now the basic examples of finite conformal modules over finite conformal algebras.

EXAMPLE 2.8a. Let $\mathfrak{g}$ be a finite-dimensional Lie algebra and let U be a finite-dimensional $\mathfrak{g}$-module. Then $\tilde{U} := U\left[t, t^{-1}\right]$ is naturally a $\tilde{\mathfrak{g}}$-module. Letting $E = \left\{u(z) := \sum_{n \in \mathbb{Z}} (ut^n) z^{-n-1} = u\delta(z - t) \,\middle|\, u \in U\right\}$, we obtain a conformal module $(\tilde{U}, E)$ over the current formal distribution Lie algebra $(\tilde{\mathfrak{g}}, F)$. Indeed, using (2.1.9), we obtain ($a \in \mathfrak{g}$, $u \in U$):

$$a(z)u(w) = (au)(w)\delta(z - w).$$

Hence the associated to $\tilde{U}$ module over the current conformal algebra $\mathrm{Cur}\,\mathfrak{g}$ is $M(\tilde{U}) = \mathbb{C}[\partial] \otimes_{\mathbb{C}} U$ defined by ($a \in \mathfrak{g}$, $u \in U$):

$$(2.8.10) \qquad a_{(0)}u = au, \quad a_{(j)}u = 0 \quad \text{for} \quad j > 0.$$

The module $M(\tilde{U})$ is finite and conformal, and it is irreducible iff U is a nontrivial irreducible $\mathfrak{g}$-module.

EXAMPLE 2.8b. Let Δ and α be complex numbers. Consider the representation of the Lie algebra $\mathrm{Vect}\,\mathbb{C}^\times$ on the following space of densities:

$$F(\Delta, \alpha) = \mathbb{C}\left[t, t^{-1}\right] e^{-\alpha t}(dt)^{1-\Delta}.$$

The action is defined as follows ($f(t) \in \mathbb{C}\left[t, t^{-1}\right]$, $g(t) \in \mathbb{C}\left[t, t^{-1}\right] e^{-\alpha t}$):

$$f(t)\partial_t \left(g(t)(dt)^{1-\Delta}\right) = \left(f(t)\partial_t g(t) + (1-\Delta)g(t)\partial_t f(t)\right)(dt)^{1-\Delta}.$$

Introduce the $F(\Delta, \alpha)$-valued formal distribution

$$m(z) := \sum_{n\in\mathbb{Z}} \left(t^n e^{-\alpha t}(dt)^{1-\Delta}\right) z^{-n-1} = \delta(t-z)e^{-\alpha t}(dt)^{1-\Delta}.$$

Recalling that $L(z) = -\delta(t-z)\partial_t$ (see Example 2.7b), and using (2.1.9), we obtain:

$$L(z)m(w) = \left((\partial_w + \alpha)m(w)\right)\delta(z-w) + \Delta m(w)\partial_w \delta(z-w).$$

Hence $(F(\Delta, \alpha), \{m(z)\})$ is a conformal module over $(\operatorname{Vect}\mathbb{C}^\times, \{L(z)\})$, and the associated finite conformal module over the Virasoro conformal algebra is $M(\Delta, \alpha) = \mathbb{C}[\partial]m$ defined by

$$(2.8.11) \qquad L_{(0)}m = (\partial + \alpha)m, \quad L_{(1)}m = \Delta m, \quad L_{(j)}m = 0 \text{ for } j > 1.$$

It is clear from the formula:

$$L_\lambda(P(\partial)m) = P(\partial + \lambda)(\partial + \alpha + \Delta\lambda)m, \quad P(\partial) \in \mathbb{C}[\partial],$$

that the module $M(\Delta, \alpha)$ is irreducible if $\Delta \neq 0$, and that $(\partial + \alpha)M(0, \alpha)$ is a nontrivial submodule of $M(0, \alpha)$.

EXAMPLE 2.8c. The formal distribution Lie algebra $\operatorname{Vect}\mathbb{C}^\times + \tilde{\mathfrak{g}}$ considered in Example 2.7c acts naturally on the space $F(\Delta, \alpha) \otimes_\mathbb{C} \tilde{U}$ (cf. Examples 2.8a, 2.8b). This is a conformal module. The associated finite conformal module over the correspoding conformal algebra $\operatorname{Conf}(\operatorname{Vect}\mathbb{C}^\times) + \operatorname{Cur}\mathfrak{g}$ is $M(\Delta, \alpha, U) = \mathbb{C}[\partial] \otimes U$ defined by ($a \in \mathfrak{g}$, $u \in U$):

$$
\begin{aligned}
&L_{(0)}u = (\partial + \alpha)u, \quad L_{(1)}u = \Delta u, \quad L_{(j)}u = 0 \quad \text{for } j > 1, \\
&a_{(0)}u = au, \quad a_{(j)}u = 0 \quad \text{for } j > 0.
\end{aligned}
$$

$(2.8.12)$

This module is irreducible iff the $\mathfrak{g}$-module U is irreducible and U is non-trivial if $\Delta = 0$.

2.9. Representation theory of finite conformal algebras

Let R be a conformal superalgebra and let $(\text{Lie}\,R, R)$ be the associated maximal formal distribution Lie superalgebra (see Section 2.7). Recall that the Lie superalgebra $\text{Lie}\,R$ admits a (even) derivation T defined by (see Remark 2.7b):

$$(2.9.1) \qquad T(a_n) = -na_{n-1}, \quad a \in R, \ n \in \mathbb{Z}.$$

It is clear from (2.7.3) that

$$(\text{Lie}\,R)_- = \mathbb{C}\text{-span of } \{a_n | a \in R, \ n \in \mathbb{Z}_+\}$$

is a subalgebra of the Lie superalgebra $\text{Lie}\,R$. It is called the *annihilation algebra*. (This subalgebra will annihilate the vacuum vector in the future vertex algebra, cf. Section 4.7, hence the name.) It is clear from (2.9.1) that $(\text{Lie}\,R)_-$ is T-invariant, hence we may form the semi-direct sum $(\text{Lie}\,R)^- = \mathbb{C}T + (\text{Lie}\,R)_-$, called the *extended annihilation algebra*.

Comparing formulas (2.7.4), (2.9.1) and definition of $\text{Lie}\,R$ with the definition of an R-module, we arrive at the following simple (but important) observation.

REMARK 2.9a. A module M over a conformal superalgebra R is the same as a module over the extended annihilation algebra $(\text{Lie}\,R)^-$. This R-module is conformal iff the following property holds:

$$(2.9.2) \qquad a_n v = 0 \quad \text{for} \quad a \in R, \ v \in M, \ n \gg 0.$$

A module over $(\text{Lie}\,R)^-$ satisfying (2.9.2) is called a *conformal* $(\text{Lie}\,R)^-$-module. A $(\text{Lie}\,R)^-$-module is called *finite* if it is finitely generated as a $\mathbb{C}[T]$-module.

REMARK 2.9b. Let M be a module over a conformal superalgebra R and let $V(M)_-$ be the $\mathbb{C}$-span of $\{v_n \in V(M) | n \in \mathbb{Z}_+\}$. This is a $(\text{Lie}\,R)^-$-submodule of the $\text{Lie}\,R$-module $V(M)$, called the *annihilation submodule*. It follows from definitions that the R-module M is isomorphic to the $(\text{Lie}\,R)^-$-module $V(M)/V(M)_-$.

Now, choose a system of generators $\{a^\alpha\}_{\alpha \in I}$ of R viewed as a $\mathbb{C}[\partial]$-module. Then we may define a descending system of subspaces

$$(2.9.3) \qquad \mathcal{L} \supset \mathcal{L}_0 \supset \mathcal{L}_1 \supset \mathcal{L}_2 \supset \ldots$$

by letting $\mathfrak{L}_n = \mathbb{C}$-span of $\{a_j^\alpha | \alpha \in I, \ j \geq n\}$. It clearly has the property

$$(2.9.4) \qquad\qquad [T, \mathfrak{L}_n] = \mathfrak{L}_{n-1} \quad \text{for} \quad n \geq 1.$$

This leads us to the following lemma.

LEMMA 2.9. [**CK2**] *Let $\mathfrak{L}$ be a Lie superalgebra with a descending system of subspaces (2.9.3) and an element T satisfying (2.9.4). Let M be an $\mathfrak{L}$-module and let*

$$M_n = \{v \in M | \mathfrak{L}_n v = 0\}.$$

(a) *Provided that U is a subspace of M_n such that $U \cap M_{n-1} = 0$ and $n \geq 1$, one has: $\mathbb{C}[T]U = \mathbb{C}[T] \otimes_{\mathbb{C}} U$. In particular, $\dim U < \infty$ if M is a finitely generated $\mathbb{C}[T]$-module.*

(b) *Suppose that $M_n \neq 0$ for some $n \in \mathbb{Z}_+$ and let N denote the minimal such n. Suppose that $N \geq 1$. Then provided that $\mathfrak{L} = \mathbb{C}T + \mathfrak{L}_0$, that $\mathfrak{L}_0$ is a subalgebra of $\mathfrak{L}$ and $[\mathfrak{L}_0, \mathfrak{L}_N] \subset \mathfrak{L}_N$ (so that $\mathfrak{L}_0 M_N \subset M_N$), the irreducibility of the $\mathfrak{L}$-module M implies*

$$(2.9.5) \qquad\qquad M = \mathbb{C}[T] \otimes_{\mathbb{C}} M_N,$$

hence the irreducibility of the $\mathfrak{L}_0$-module M_N. Conversely, if the $\mathfrak{L}_0$-module M_N is irreducible and has no non-zero vectors annihilated by $\mathfrak{L}_{N-1}$, then the $\mathfrak{L}$-module (2.9.5) is irreducible.

PROOF. Let L_a and R_a denote the operator of left and right multiplication by an element a of an associative algebra A. Using $R_a = L_a - \operatorname{ad} a$ and the binomial formula, we get the following well-known formula in A:

$$(2.9.6) \qquad\qquad ga^k = \sum_{j=0}^{k} \binom{k}{j} a^{k-j}(-\operatorname{ad} a)^j g, \quad a, g \in A.$$

Let $\{v_i\}_{i \in I}$ be a $\mathbb{C}$-linearly independent set of vectors in U generating the $\mathbb{C}[T]$-module $\mathbb{C}[T]U$. Suppose that $\sum_i p_i(T)v_i = 0$, where $p_i(T) \in \mathbb{C}[T]$, and not all $p_i(T) = 0$. Let m be the maximal degree of the $p_i(T)$'s. We write $p_i(T) = \sum_{j=0}^{m} c_{ij}T^j$, $c_{ij} \in \mathbb{C}$, so that we have $c_{im} \neq 0$ for some i. Using (2.9.6) and (2.9.4), we have, since $n \geq 1$:

$$\mathcal{L}_{n+m-1}T^k = \sum_{j=0}^{k} \binom{k}{j} T^{k-j} (\operatorname{ad} T)^j (\mathcal{L}_{n+m-1})$$

$$= \sum_{j=0}^{k} \binom{k}{j} T^{k-j} \mathcal{L}_{n+m-1-j}.$$

We have therefore

$$0 = \mathcal{L}_{n+m-1} \sum_i p_i(T)v_i = \sum_i c_{im}\mathcal{L}_{n-1}v_i = \mathcal{L}_{n-1}\left(\sum_i c_{im}v_i \right).$$

Since $\sum_i c_{im}v_i \neq 0$, we arrive at a contradiction, proving (a).

Under the assumptions of (b), if M is an irreducible $\mathcal{L}$-module, then $M = \mathbb{C}[T]M_N$, hence by (a), (2.9.5) holds and M_N must be an irreducible $\mathcal{L}_0$-module. Conversely, if the $\mathcal{L}_0$-module M_N is irreducible, but the $\mathcal{L}$-module (2.9.5) is reducible, then a non-trivial quotient of the latter would contradict (a) for $U = M_N$. $\square$

Now it is easy to classify all finite conformal irreducible modules over the most important finite conformal algebras. An R-module M is called *trivial* if $a^M_{(n)}m = 0$ for all $a \in R$, $m \in M$, $n \in \mathbb{Z}_+$.

THEOREM 2.9. *Let R be a conformal algebra of one of the three types described by Examples 2.7a-2.7c. Then a complete list of non-trivial conformal finite irreducible R-modules M is as follows.*

(a) If R is the current conformal algebra $\operatorname{Cur}\mathfrak{g}$, where $\mathfrak{g}$ is a finite-dimensional semisimple Lie algebra, then $M \simeq M(\tilde{U})$, where U is a non-trivial finite-dimensional irreducible $\mathfrak{g}$-module (see Example 2.8a).

(b) If R is the Virasoro conformal algebra, then $M \simeq M(\Delta, \alpha)$ with $\Delta \neq 0$ (see Example 2.8b).

(c) If R is the semi-direct sum of the Virasoro conformal algebra and the current conformal algebra $\operatorname{Cur}\mathfrak{g}$, where $\mathfrak{g}$ is a finite-dimensional Lie algebra, then $M \simeq M(\Delta, \alpha, U)$, where U is a finite-dimensional irreducible $\mathfrak{g}$-module which must be non-trivial if $\Delta = 0$ (see Example 2.8c).

PROOF. Let $R = \operatorname{Cur}\mathfrak{g}$. Then we have:

$$\mathcal{L} := (\operatorname{Lie} R)^- = \mathbb{C}T + \mathfrak{g}[t], \quad T = -\partial_t,$$

with the filtration $\mathcal{L}_n = \mathfrak{g}[t]t^n$, $n \in \mathbb{Z}_+$. Let M be an irreducible R-module. Then, by Remark 2.9a, it is a conformal $\mathcal{L}$-module and we may apply Lemma 2.9. If $N \geq 1$, we have (2.9.5), where M_N is an irreducible $\mathcal{L}_0/\mathcal{L}_N$-module. If, in addition, M is finite module, then $\dim_{\mathbb{C}} M_N < \infty$, and we may apply a well-known result from Lie algebra theory (see e.g. [**Se**]) to show that M_N is an irreducible $\mathfrak{g}[t]$-module with trivial action of $\mathfrak{g}[t]t$.

If $N = 0$, then M_N is a trivial $\mathfrak{g}[t]$-module, hence an $\mathcal{L}$-submodule of M, hence $M = M_N$ and M is a trivial R-module. This proves (a).

Let Vir be the Virasoro conformal algebra. Then we have:

$$(\mathrm{Lie}\,\mathrm{Vir})^- = \mathbb{C}T + \mathrm{Vect}\,\mathbb{C},$$

where $\mathrm{Vect}\,\mathbb{C} = \oplus_{n \in \mathbb{Z}_+} \mathbb{C}t^n \partial_t$ and T acts on it as $-\mathrm{ad}\,\partial_t$. It follows that $(\mathrm{Lie}\,\mathrm{Vir})^-$ is a direct sum (as ideals) of the commutative Lie algebra $\mathbb{C}(T + \partial_t)$ and the Lie algebra $\mathcal{L} := \mathrm{Vect}\,\mathbb{C}$. Let M be an irreducible Vir-module. By Remark 2.9a, it is an irreducible $(\mathrm{Lie}\,\mathrm{Vir})^-$-module. Hence $T + \partial_t$ acts as a scalar, which we denote by α and $\mathcal{L}$ acts irreducibly on M. Define the following filtration on $\mathcal{L} : \mathcal{L}_n = \oplus_{j \geq n} \mathbb{C}t^{j+1} \partial_t$ and apply Lemma 2.9. If $N \geq 1$, we argue in the same way as in the case (a) to show (2.9.5) with M_N irreducible and to show that $M \simeq M(\Delta, \alpha)$ with $\Delta \neq 0$ if M is finite. If $N = 0$, then it is easy to see that M is the 1-dimensional trivial $\mathcal{L}$-module, proving (b).

The proof of (c) is similar. $\square$

A conformal $(\mathfrak{g}, F)$-module (V, E) is called *finite* if $\bar{E}$ is a finitely-generated $\mathbb{C}[\partial]$-module and is called *irreducible* if it contains only irregular non-zero submodules. (As before a submodule $I \subset V$ is called irregular if it does not contain all coefficients of a non-zero formal distribution from $\bar{E}$.) A conformal $(\mathfrak{g}, F)$-module (V, E) is called *trivial* if $\mathfrak{g}V = 0$. In view of the discussion in Section 2.8, Theorem 2.9 is equivalent to the following corollary.

COROLLARY 2.9. *All non-trivial finite irreducible conformal modules over the loop algebra $\tilde{\mathfrak{g}}$, where $\mathfrak{g}$ is a finite-dimensional semisimple Lie algebra, over the Lie algebra $\mathrm{Vect}\,\mathbb{C}^\times$, and over their semi-direct sum are respectively: quotients of loop modules $\tilde{U}$, where U is a non-trivial finite-dimensional irreducible $\mathfrak{g}$-module; the* *density modules $F(\Delta, \alpha)$, where $\Delta \neq 0$; and the modules $\tilde{U}e^{-\alpha t}(dt)^{1-\Delta}$, where U is a finite-dimensional irreducible $\mathfrak{g}$-module which must be non-trivial if $\Delta = 0$.*

In conclusion of this section, we show how one uses Lemma 2.9 in order to prove an analogue of classical Lie theorem on representations of solvable Lie algebras.

CONFORMAL ANALOGUE OF LIE THEOREM [**DK**]. *Let M be a finite conformal module over a finite solvable conformal algebra R. Then there exists a common eigenvector (with eigenvalues in $\mathbb{C}$) of all the operators $a^M_{(n)}$, where $a \in R$, $n \in \mathbb{Z}_+$.*

PROOF. We prove the theorem by induction on the lexicographically ordered pair (rank R, dim tor R) of non-negative integers.

Let $S \subset R$ be the last non-zero member of the derived series of R. Then S is commutative and $R_{(j)}S \subset S$ for all $j \in \mathbb{Z}_+$. Hence we have a representation of R/S in S. By the inductive assumption applied to the conformal algebra R/S, we may deduce that there exists a non-zero element $b \in S$ such that:

$$(2.9.7) \qquad\qquad R_{(j)}b \in \mathbb{C}b \quad \text{for} \quad j \in \mathbb{Z}_+.$$

Consider the Lie algebras of operators on M

$$\mathfrak{g} = (\operatorname{Lie} R)^M_- = \sum_{a \in R, j \in \mathbb{Z}_+} \mathbb{C}a^M_{(j)}$$

and

$$\mathfrak{b} = \mathbb{C}\partial^M + \sum_{j \in \mathbb{Z}_+} \mathbb{C}b^M_{(j)}$$

with the filtration

$$\mathfrak{b}_n = \sum_{j \geq n} \mathbb{C}b^M_{(j)}.$$

Recalling that $\left[a^M_{(m)}, b^M_{(n)}\right] = \sum_{j \geq 0} \binom{m}{j} \left(a_{(j)}b\right)^M_{(m+n-j)}$, we see, using (2.9.7), that

$$(2.9.8) \qquad\qquad [\mathfrak{g}, \mathfrak{b}_n] \subset \mathfrak{b}_n.$$

Let $M_n = \{v \in M \,|\, \mathfrak{b}_n v = 0\}$, and let N be the minimal $n \in \mathbb{Z}_+$ such that $M_n \neq 0$.

Case 1. $N = 0$. Then, due to (2.9.8), M_0 is a non-zero R-submodule, hence a $R/\mathbb{C}[\partial]b$ submodule, and we may apply the inductive assumption.

Case 2. $N \geq 1$. Then, by Lemma 2.9 applied to the $\mathfrak{b}$-module M, $\dim_{\mathbb{C}} M_N < \infty$. By (2.9.8), $\mathfrak{g}M_N \subset M_N$, and since $\mathfrak{g}$ is a solvable Lie algebra, we may apply the classical Lie theorem (see e.g. [**Se**]) to find an eigenvector for $\mathfrak{g}$ in M_N. $\square$

2.10. Associative conformal algebras and the general conformal algebra

Some of the main examples of Lie algebras are associative algebras with the Lie bracket. Here we discuss a similar construction in the "conformal" framework.

Let A be an associative algebra over $\mathbb{C}$. Two A-valued formal distributions $a(z)$ and $b(w)$ are called *local* (or form a *local pair*) if the formal distribution $a(z)b(w)$ is local. Due to Corollary 2.2, we have the expansion into a finite sum for any local pair of A-valued formal distributions:

$$(2.10.1) \qquad a(z)b(w) = \sum_{j \in \mathbb{Z}_+} (a(w)_j b(w))\, \partial_w^{(j)} \delta(z - w),$$

where

$$(2.10.2) \qquad a(w)_j b(w) = \mathrm{Res}_z\, (z - w)^j a(z)b(w).$$

Suppose that the algebra A is spanned by coefficients of a family F of pairwise local formal distributions. Then (A, F) is called a *formal distribution associative algebra*.

As before, we consider the closure $\bar{F}$ of F, which is the minimal $\mathbb{C}[\partial]$-module containing F and closed under all products (2.10.2). All pairs from $\bar{F}$ are local due to an "associative" analogue of Lemma 2.8 (which is easy to prove).

As before, the properties of the products $a(w)_j b(w)$ on $\bar{F}$ are neatly described in terms of the λ-product

$$a(w)_\lambda b(w) = \sum_{j \in \mathbb{Z}_+} \lambda^{(j)} a(w)_j b(w).$$

As in the Lie algebra case, this leads us to the notion of an *associative conformal algebra*. This is a $\mathbb{C}[\partial]$-module R endowed with the λ-product $R \otimes_{\mathbb{C}} R \to \mathbb{C}[\lambda] \otimes_{\mathbb{C}} R$, denoted by $a_\lambda b$, satisfying the following axioms:

$(\mathrm{A}1)_\lambda \qquad\qquad (\partial a)_\lambda b = -\lambda a_\lambda b, \quad a_\lambda \partial b = (\partial + \lambda)(a_\lambda b),$

$(\mathrm{A}2)_\lambda \qquad\qquad a_\lambda (b_\mu c) = (a_\lambda b)_{\lambda+\mu} c.$

Of course, writing $a_\lambda b = \sum_{j \in \mathbb{Z}_+} \lambda^{(j)} a_j b$, one may write equivalent axioms for the products $a_j b$.

As in the Lie algebra case, we have an associative conformal algebra $\bar{F}$ associated to any formal distribution associative algebra (A, F). Conversely, introducing the affinization $\tilde{R}$ of an associative conformal algebra R, we may construct the formal distribution associative algebra $\operatorname{Ass} R = \tilde{R}/\partial\tilde{R}$, in the same way as we did in Section 2.7 for Lie algebras. As in the Lie algebra case, this establishes a bijective correspondence between associative conformal algebras R and families of formal distribution associative algebras obtained from $\operatorname{Ass} R$ as quotients by irregular ideals. Similarly, one defines conformal modules over formal distribution associative algebras and establishes their correspondence to conformal modules over associative conformal algebras as in Section 2.8. A *conformal module* over an associative conformal algebra R is a $\mathbb{C}[\partial]$-module M endowed with a $\mathbb{C}$-linear map $R \to \mathbb{C}[\lambda] \otimes_\mathbb{C} M$, denoted by $a \mapsto a_\lambda^M$, satisfying the properties:

$$(\partial a)_\lambda^M = [\partial^M, a_\lambda^M] = -\lambda a_\lambda^M, \quad a \in R,$$

$$a_\lambda^M b_\mu^M = (a_\lambda b)_{\lambda+\mu}^M, \quad a, b \in R.$$

REMARK 2.10a. Let (A, F) be a formal distribution associative algebra. Then (A^{op}, F), where A^{op} is the associative algebra with the opposite multiplication, is a formal distribution associative algebra as well. Translating into the language of associative conformal algebras, we see, using Proposition 2.3(b), that, given an associative conformal algebra R with λ-product $a_\lambda b$, its opposite associative conformal algebra R^{op} has λ-product $b_{-\lambda-\partial}a$. In particular, the λ-bracket

$$(2.10.3) \qquad [a_\lambda b] = a_\lambda b - b_{-\lambda-\partial}a$$

makes R a conformal algebra (satisfying axioms $(\mathrm{C1})_\lambda$-$(\mathrm{C3})_\lambda$).

REMARK 2.10b. An *associative conformal superalgebra* R is simply a $\mathbb{Z}/2\mathbb{Z}$-graded associative conformal algebra. The λ-bracket (cf. Section 2.3)

$$[a_\lambda b] = a_\lambda b - p(a, b) b_{-\lambda-\partial}a$$

turns R into a conformal superalgebra.

An associative conformal algebra A is called *commutative* if

$$a_\lambda b = b_{-\lambda-\partial}a.$$

Obviously, commutative associative conformal algebras correspond to formal distribution commutative associative algebras. It would be very interesting to develop an algebraic geometry based on commutative associative conformal algebras.

Now we turn to examples.

EXAMPLE 2.10a. If A is an arbitrary associative algebra, then the corresponding current algebra $A\left[t, t^{-1}\right]$ is a formal distribution associative algebra with the family $F = \left\{a(z) = \sum_n (at^n)z^{-n-1} \,\middle|\, a \in A\right\}$ of local formal distributions. Indeed, we have:

$$a(z)b(w) = (ab)(w)\delta(z - w).$$

The corresponding associative conformal algebra is $R = \mathbb{C}[\partial] \otimes_\mathbb{C} A$ with λ-product

$$a_\lambda b = ab, \quad a, b \in A.$$

A much more interesting example is the following.

EXAMPLE 2.10b. Let $\operatorname{Diff} \mathbb{C}^\times$ be the associative algebra of regular differential operators on $\mathbb{C}^\times$. It has a basis $t^j \partial_t^m$, $j \in \mathbb{Z}$, $m \in \mathbb{Z}_+$. Introduce the formal distributions $(m \in \mathbb{Z}_+)$:

$$J^m(z) = \sum_{j \in \mathbb{Z}} t^j (-\partial_t)^m z^{-j-1} = \delta(t - z)(-\partial_t)^m.$$

Using (2.1.10) (for $n = 0$), we obtain:

$$J^m(z)J^n(w) = \sum_{j=0}^{m} \sum_{i=0}^{j} \binom{m}{j}\binom{j}{i} \partial_w^{j-i} J^{m+n-j}(w)\partial_w^i \delta(z - w).$$

It follows that the family $F = \{J^m(z) | m \in \mathbb{Z}_+\}$ consists of pairwise local formal distributions with products:

$$(2.10.4) \qquad J^m(w)_i J^n(w) = \sum_{j=i}^{m} i!\binom{m}{j}\binom{j}{i} \partial_w^{j-i} J^{m+n-j}(w).$$

Hence $(\operatorname{Diff}\mathbb{C}^\times, F)$ is a formal distribution associative algebra. The corresponding associative conformal algebra is

$$\operatorname{Conf}(\operatorname{Diff}\mathbb{C}^\times, F) = \oplus_{m \in \mathbb{Z}_+} \mathbb{C}[\partial] J^m$$

with the λ-product, derived from (2.10.4) being as follows ($m, n \in \mathbb{Z}_+$):

$$(2.10.5) \qquad J^m{}_\lambda J^n = \sum_{j=0}^{m} \binom{m}{j} (\lambda + \partial)^j J^{m+n-j}.$$

Consider the obvious representation of the algebra $\mathrm{Diff}\,\mathbb{C}^\times$ on the space $\mathbb{C}\left[t, t^{-1}\right]$. Letting $v(z) = \sum_{n \in \mathbb{Z}} t^n z^{-n-1} = \delta(z - t)$, we obtain, using (2.1.10):

$$(2.10.6) \qquad J^m(z)v(w) = \sum_{j=0}^{m} m(m-1)\ldots(m-j+1)\partial_w^{m-j} v(w)\partial_w^{(j)}\delta(z - w).$$

Hence $\left(\mathbb{C}\left[t, t^{-1}\right], \{v(w)\}\right)$ is a conformal module over $(\mathrm{Diff}\,\mathbb{C}^\times, F)$. The associated conformal module over $\mathrm{Conf}(\mathrm{Diff}\,\mathbb{C}^\times, F)$ is $\mathbb{C}[\partial]v$ with the λ-action obtained from (2.10.6) to be given by

$$(2.10.7) \qquad J_\lambda^m v = (\lambda + \partial)^m v, \quad m \in \mathbb{Z}_+.$$

(A simpler way to derive formulas (2.10.5) and (2.10.7) is to use Lemma 2.2.)

A matrix generalization of this example is also important.

EXAMPLE 2.10c. The associative algebra

$$\mathrm{Diff}_N \mathbb{C}^\times = \left(\mathrm{Diff}\,\mathbb{C}^\times\right) \otimes_{\mathbb{C}} \mathrm{Mat}_N \mathbb{C}$$

of all $N \times N$ matrix valued regular differential operators on $\mathbb{C}^\times$ is a formal distribution associative algebra with the family of pairwise local formal distributions

$$F = \left\{ J_A^m(z) = J^m(z) \otimes A \mid m \in \mathbb{Z}_+, A \in \mathrm{Mat}_N \mathbb{C}\right\}.$$

The associated associative conformal algebra is

$$\mathrm{Conf}\left(\mathrm{Diff}_N \mathbb{C}^\times, F\right) = \oplus_{m \in \mathbb{Z}_+}\mathbb{C}[\partial]\left(J^m \otimes_{\mathbb{C}} \mathrm{Mat}_N \mathbb{C}\right)$$

with λ-products

$$(2.10.8) \qquad J_A^m{}_\lambda J_B^n = \sum_{j=0}^{m} \binom{m}{j}(\lambda + \partial)^j J_{AB}^{m+n-j}.$$

The obvious representation of $\mathrm{Diff}_N \mathbb{C}^\times$ on the space $\mathbb{C}\left[t, t^{-1}\right] \otimes \mathbb{C}^N$ is an irreducible conformal module with the family $E = \left\{v_a(w) = v(w) \otimes a \mid a \in \mathbb{C}^N\right\}$. The associated (conformal) module over $\mathrm{Conf}(\mathrm{Diff}_N \mathbb{C}^\times, F)$ is $\mathbb{C}[\partial] \otimes_{\mathbb{C}} \mathbb{C}^N$ with the λ-action

$$(2.10.9) \qquad J_A^m{}_\lambda v = (\lambda + \partial)^m Av, \quad m \in \mathbb{Z}_+, \, v \in \mathbb{C}^N.$$

A more conceptual understanding of Example 2.10c is given by Proposition 2.10 below.

DEFINITION 2.10. Let U and V be two $\mathbb{C}[\partial]$-modules. A *conformal linear map* from U to V is a $\mathbb{C}$-linear map $a : U \to \mathbb{C}[\lambda] \otimes_{\mathbb{C}} V$, denoted by $a_\lambda : U \to V$, such that

$$[\partial, a_\lambda] = -\lambda a_\lambda.$$

(This equation means: $\partial^V a_\lambda - a_\lambda \partial^U = -\lambda a_\lambda$.) Denote the vector space (over $\mathbb{C}$) of all such maps by $\mathrm{Chom}(U, V)$. It has a canonical structure of a $\mathbb{C}[\partial]$-module:

$$(\partial a)_\lambda = -\lambda a_\lambda.$$

REMARK 2.10c. Let U and V be modules over a conformal algebra R. Then the $\mathbb{C}[\partial]$-module $H := \mathrm{Chom}(U, V)$ carries an R-module structure defined by ($a \in R, \varphi \in H, u \in U$):

$$\left(a_\lambda^H \varphi\right)_\mu u = a_\lambda^V \left(\varphi_{\mu-\lambda} u\right) - \varphi_{\mu-\lambda} \left(a_\lambda^U u\right).$$

Hence one may define the contragredient conformal R-module $U^* = \mathrm{Chom}(U, \mathbb{C})$, where $\mathbb{C}$ is the trivial R-module and $\mathbb{C}[\partial]$-module, and the tensor product of R-modules: $U \otimes V = \mathrm{Chom}(U^*, V)$. It is easy to see that the R-module $\mathrm{Chom}(U, V)$ is conformal iff both U and V are finite conformal R-modules.

In the special case $U = V$ we let $\mathrm{Cend}\, V = \mathrm{Chom}(V, V)$. Provided that V is a finite $\mathbb{C}[\partial]$-module, the $\mathbb{C}[\partial]$-module $\mathrm{Cend}\, V$ has a canonical structure of an associative conformal algebra defined for $a, b \in \mathrm{Cend}\, V$ by

$$(2.10.10) \qquad (a_\lambda b)_\mu^V v = a_\lambda^V \left(b_{\mu-\lambda}^V v\right), \quad v \in V.$$

Indeed, axiom $(A1)_\lambda$ is immediate, while axiom $(A2)_\lambda$ is obtained from $(2.10.10)$ by replacing μ by $\mu + \lambda$. Finally, it is easy to show that $a_\lambda b$ depends polynomially on λ using that V is a finite $\mathbb{C}[\partial]$-module.

REMARK 2.10d. By the very definition, a structure of a conformal module over an associative conformal algebra R in a finite $\mathbb{C}[\partial]$-module V is the same as a homomorphism of R to the associative conformal algebra $\mathrm{Cend}\, V$.

The λ-bracket (2.10.3) on $\operatorname{Cend} V$, where V is a finite $\mathbb{C}[\partial]$-module, makes it a conformal algebra, which we denote by $\operatorname{gc} V$ and call the *general conformal algebra*. The second term of the bracket (2.10.3) can be simplified:

$$- (b_{-\lambda-\partial}a)^V_\mu\, v = - \sum_{n\geq 0} \left((-\lambda - \partial)^{(n)}(b_n a)\right)^V_\mu v = - (b_{\mu-\lambda}a)^V_\mu\, v = -b^V_{\mu-\lambda}\left(a^V_\lambda v\right).$$

Thus the λ-bracket of $\operatorname{gc} V$ looks as follows:

$$(2.10.11) \qquad\qquad [a_\lambda b]^V_\mu\, v = \left[a^V_\lambda, b^V_{\mu-\lambda}\right] v.$$

REMARK 2.10e. Formula (2.10.11) shows that a structure of a conformal module over a conformal algebra R in a finite $\mathbb{C}[\partial]$-module V is the same as a homomorphism of R to the conformal algebra $\operatorname{gc} V$.

For a positive integer N we let $\operatorname{Cend}_N = \operatorname{Cend} \mathbb{C}[\partial]^N$, $\operatorname{gc}_N = \operatorname{gc} \mathbb{C}[\partial]^N$ (where $\mathbb{C}[\partial]^N$ is the free $\mathbb{C}[\partial]$-module of rank N). Recall that we have a representation of the associative conformal algebra $\operatorname{Conf}(\operatorname{Diff} \mathbb{C}^\times)$ in $\mathbb{C}[\partial]^N$ defined by (2.10.9). By Remark 2.10d this gives us a homomorphism $\varphi : \operatorname{Conf}\left(\operatorname{Diff}_N \mathbb{C}^\times\right) \to \operatorname{Cend}_N$ of associative conformal algebras. Likewise, by Remark 2.10e we get a conformal algebra homomorphism $\varphi_- : \operatorname{Conf}\left(\operatorname{Diff}_N^- \mathbb{C}^\times, F\right) \to \operatorname{gc}_N$, where $\operatorname{Diff}_N^- \mathbb{C}^\times$ stands for $\operatorname{Diff}_N \mathbb{C}^\times$ with the usual Lie bracket.

PROPOSITION 2.10. [**DK**] *The homomorphisms φ and φ_- are isomorphisms.*

PROOF. We have by (2.10.9):

$$\left(\partial^k J_A^m\right)_\lambda v = (-\lambda)^k(\lambda + \partial)^m Av, \quad k, m \in \mathbb{Z}_+,\ v \in \mathbb{C}^N.$$

This formula shows that φ and φ_- are injective. The same formula shows that φ and φ_- are surjective since a conformal linear map is determined by its values on a set of the generators of a $\mathbb{C}[\partial]$-module, but the polynomials $\lambda^k(\lambda + \partial)^m v$ $(k, m \in \mathbb{Z}_+,\ v \in \mathbb{C}^N)$ span over $\mathbb{C}$ the space $\mathbb{C}[\lambda, \partial] \otimes \mathbb{C}^N$. $\qquad\square$

REMARK 2.10f. The associative conformal algebra Cend_N and the general conformal algebra gc_N are interesting examples of simple algebras which are not finite (but have finite Gelfand-Kirillov dimension). It is an interesting open problem to classify such algebras. A related open problem is to classify infinite subalgebras of Cend_N and gc_N which act irreducibly on $\mathbb{C}[\partial]^N$. (For a classification of such finite algebras see [**DK**].)

2.11. Cohomology of conformal algebras

This section is an exposition of some of the results of the paper [**BKV**]. (A generalization to the super case is straightforward by making use of the usual sign rule.)

DEFINITION 2.11a. An *n-cochain* $(n \in \mathbb{Z}_+)$ *of a conformal algebra R with coefficients in an R-module over it* is a $\mathbb{C}$-linear map

$$\gamma : R^{\otimes n} \to M[\lambda_1, \ldots, \lambda_n], \quad a_1 \otimes \cdots \otimes a_n \mapsto \gamma_{\lambda_1,\ldots,\lambda_n}(a_1, \ldots, a_n),$$

where $M[\lambda_1, \ldots, \lambda_n]$ denotes the space of polynomials with coefficients in M, satisfying the following conditions:

(2.11.1) $\gamma_{\lambda_1,\ldots,\lambda_n}(a_1, \ldots, \partial a_i, \ldots, a_n) = -\lambda_i \gamma_{\lambda_1,\ldots,\lambda_n}(a_1, \ldots, a_i, \ldots, a_n),$

(2.11.2) $\quad\quad\quad$ γ is skew-symmetric with respect to simultaneous permutations of a_i's and λ_i's.

We let $R^{\otimes 0} = \mathbb{C}$, so that a 0-cochain γ is an element of M and $(d\gamma)_\lambda(a) = a_\lambda \gamma$. Sometimes, when the module M is not conformal, one may consider formal power series instead of polynomials in this definition.

We define a differential d of an n-cochain γ as follows:

$$(d\gamma)_{\lambda_1,\ldots,\lambda_{n+1}}(a_1, \ldots, a_{n+1})$$
$$= \sum_{i=1}^{n+1} (-1)^{i+1} a_{i\lambda_i} \gamma_{\lambda_1,\ldots,\widehat{\lambda_i},\ldots,\lambda_{n+1}}(a_1, \ldots, \widehat{a_i}, \ldots, a_{n+1})$$
$$+ \sum_{\substack{i,j=1 \\ i<j}}^{n+1} (-1)^{i+j} \gamma_{\lambda_i+\lambda_j,\lambda_1,\ldots,\widehat{\lambda_i},\ldots,\widehat{\lambda_j},\ldots,\lambda_{n+1}}([a_{i\lambda_i} a_j], a_1, \ldots, \widehat{a_i}, \ldots, \widehat{a_j}, \ldots, a_{n+1}),$$

where γ is extended linearly over the polynomials in λ_i.

REMARK 2.11a. Property (2.11.1) implies the following relation for an n-cochain γ:

$$\gamma_{\lambda+\mu,\lambda_1,\ldots,\lambda_{n-1}}([a_\lambda b], a_1, \ldots, a_{n-1}) = \gamma_{\lambda+\mu,\lambda_1,\ldots,\lambda_{n-1}}([a_{-\partial-\mu} b], a_1, \ldots, a_{n-1}).$$

LEMMA 2.11. (a) *The operator d preserves the space of cochains.*

(b) $d^2 = 0$.

PROOF. (a) Property (2.11.1) obviously holds for $d\gamma$ if it holds for γ. The only non-trivial point in checking (2.11.2) of $d\gamma$ amounts to the equation

$$\gamma_{\lambda+\mu,\lambda_1,\ldots,\lambda_{n-1}}([a_\lambda b], a_1, \ldots, a_{n-1}) = -\gamma_{\lambda+\mu,\lambda_1,\ldots,\lambda_{n-1}}([b_\mu a], a_1, \ldots, a_{n-1}),$$

which follows from Remark 2.11a and the skew-symmetry $(C2)_\lambda$ of $[a_\lambda b]$.

(b) We have for an n-cochain γ

$$(d^2\gamma)_{\lambda_1,\ldots,\lambda_{n+2}}(a_1,\ldots,a_{n+2})$$

$$= \sum_{i=1}^{n+2}(-1)^{i+1}a_{i\lambda_i}(d\gamma)_{\lambda_1,\ldots,\widehat{\lambda}_i,\ldots,\lambda_{n+2}}(a_1,\ldots,\widehat{a}_i,\ldots,a_{n+2})$$

$$+ \sum_{\substack{i,j=1\\i<j}}^{n+2}(-1)^{i+j}(d\gamma)_{\lambda_i+\lambda_j,\lambda_1,\ldots,\widehat{\lambda}_{i,j},\ldots,\lambda_{n+2}}([a_{i\lambda_i}a_j],a_1,\ldots,\widehat{a}_{i,j},\ldots,a_{n+2})$$

$$= \sum_{\substack{i,j=1\\i\neq j}}^{n+2}(-1)^{i+j}\,\mathrm{sign}\{j,i\}\,a_{i\lambda_i}\big(a_{j\lambda_j}\gamma_{\lambda_1,\ldots,\widehat{\lambda}_{i,j},\ldots,\lambda_{n+2}}(a_1,\ldots,\widehat{a}_{i,j},\ldots,a_{n+2})\big)$$

$$+ \sum_{\substack{i,j,k=1\\j<k,i\neq j,k}}^{n+2}(-1)^{i+j+k+1}\,\mathrm{sign}\{j,k,i\}\,a_{i\lambda_i}\gamma_{\lambda_j+\lambda_k,\lambda_1,\ldots,\widehat{\lambda}_{i,j,k},\ldots,\lambda_{n+2}}$$

$$([a_{j\lambda_j}a_k],a_1,\ldots,\widehat{a}_{i,j,k},\ldots,a_{n+2})$$

$$+ \sum_{\substack{i,j,k=1\\i<j,k\neq i,j}}^{n+2}(-1)^{i+j+k}\,\mathrm{sign}\{k,i,j\}a_{k\lambda_k}\gamma_{\lambda_i+\lambda_j,\lambda_1,\ldots,\widehat{\lambda}_{i,j,k},\ldots,\lambda_{n+2}}$$

$$([a_{i\lambda_i}a_j],a_1,\ldots,\widehat{a}_{i,j,k},\ldots,a_{n+2})$$

$$+ \sum_{\substack{i,j=1\\i<j}}^{n+2}(-1)^{i+j}[a_{i\lambda_i}a_j]_{\lambda_i+\lambda_j}\gamma_{\lambda_1,\ldots,\widehat{\lambda}_{i,j},\ldots,\lambda_{n+2}}(a_1,\ldots,\widehat{a}_{i,j},\ldots,a_{n+2})$$

$$+ \sum_{\substack{i,j,k,l=1\\i<j\neq k<l}}^{n+2}(-1)^{i+j+k+l}\,\mathrm{sign}\{i,j\neq k,l\}\,\gamma_{\lambda_k+\lambda_l,\lambda_i+\lambda_j,\lambda_1,\ldots,\widehat{\lambda}_{i,j,k,l},\ldots,\lambda_{n+2}}$$

$$([a_{k\lambda_k}a_l],[a_{i\lambda_i}a_j],a_1,\ldots,\widehat{a}_{i,j,k,l},\ldots,a_{n+2})$$

$$+ \sum_{\substack{i,j,k=1\\i<j,k\neq i,j}}^{n+2}(-1)^{i+j+k+1}\,\mathrm{sign}\{i,j,k\}\,\gamma_{\lambda_i+\lambda_j+\lambda_k,\lambda_1,\ldots,\widehat{\lambda}_{i,j,k},\ldots,\lambda_{n+2}}$$

$$([[a_{i\lambda_i}a_j]_{\lambda_i+\lambda_j}a_k],a_1,\ldots,\widehat{a}_{i,j,k},\ldots,a_{n+2}),$$

where $\mathrm{sign}\{i_1,\ldots,i_p\}$ is the sign of the permutation putting the indices in the increasing order and $\widehat{a}_{i,j,\ldots}$ means that $a_i,a_j,\ldots$ are omitted. Notice that each term in the summation over i, j, k, l is skew-symmetric with respect to the permutation $\begin{pmatrix} i & j & k & l \\ k & l & i & j \end{pmatrix}$. Therefore, the terms of that summation will cancel pairwise. The first and the forth summations cancel each other, because M is a conformal algebra module:

$$-a_{i\lambda_i}(a_{j\lambda_j}m) + a_{j\lambda_j}(a_{i\lambda_i}m) + [a_{i\lambda_i}a_j]_{\lambda_i+\lambda_j}m = 0.$$

The second summation becomes equal to the third one after the substitution (ikj), except they differ by a sign. Thus, they cancel each other, as well. Finally, the sixth summation can be rewritten as summation over $i < j < k$ of the sum of three permutations of the initial summand. Precisely, in the first entry of γ, we will have

$$[[a_{i\lambda_i}a_j]_{\lambda_i+\lambda_j}a_k] - [[a_{i\lambda_i}a_k]_{\lambda_i+\lambda_k}a_j] + [[a_{j\lambda_j}a_k]_{\lambda_j+\lambda_k}a_i].$$

Using Remark 2.11a, we can transform this sum inside γ into

$$[[a_{i\lambda_i}a_j]_{\lambda_i+\lambda_j}a_k] - [[a_{i\lambda_i}a_k]_{-\partial-\lambda_j}a_j] + [[a_{j\lambda_j}a_k]_{-\partial-\lambda_i}a_i],$$

which vanishes by the Jacobi identity $(C3)_\lambda$ and skew-symmetry $(C2)_\lambda$ in R. Thus, we see that all of the terms in $d^2\gamma$ cancel. $\square$

Thus the cochains of a conformal algebra R with coefficients in an R-module M form a complex:

$$\tilde{C}(R, M) = \bigoplus_{n\in\mathbb{Z}_+} \tilde{C}^n(R, M),$$

where $\tilde{C}^n(R, M)$ denotes the space of all n-cochains. This complex is called the *basic complex* for the R-module M. This is not yet the complex defining the right cohomology of a conformal algebra: we need to consider a certain quotient complex.

Define the structure of a $\mathbb{C}[\partial]$-module on $\tilde{C}(R, M)$ by letting

$$(2.11.3) \qquad (\partial\gamma)_{\lambda_1,\ldots,\lambda_n}(a_1,\ldots,a_n) = \left(\partial^M + \sum_{i=1}^n \lambda_i\right)\gamma_{\lambda_1,\ldots,\lambda_n}(a_1,\ldots,a_n),$$

where ∂^M denotes the action of ∂ on M.

REMARK 2.11b. $d\partial = \partial d$, and therefore the graded subspace $\partial\tilde{C}(R, M)$ of $\tilde{C}(R, M)$ is a subcomplex. Indeed, the first summation in the differential transforms the factor $\partial^M + \sum_{i=1}^n \lambda_i$ into $\partial^M + \sum_{i=1}^{n+1} \lambda_i$, because of the properties $(C1)_\lambda$ and

$(C1')_\lambda$ of the λ-bracket. The second summation does the same for more obvious reasons.

Define the quotient complex

$$C(R, M) = \tilde{C}(R, M)/\partial\tilde{C}(R, M) = \bigoplus_{n \in \mathbb{Z}_+} C^n(R, M),$$

called the *reduced complex*.

DEFINITION 2.11b. The *basic cohomology* $\tilde{H}(R, M) = \bigoplus_{n \in \mathbb{Z}_+} \tilde{H}^n(R, M)$ is the cohomology of the basic complex $\tilde{C}(R, M)$. The *reduced cohomology* $H(R, M) = \bigoplus_{n \in \mathbb{Z}_+} H^n(R, M)$ of a conformal algebra R with coefficients in a module M is the cohomology of the *reduced complex* $C(R, M)$.

REMARK 2.11c. The exact sequence $0 \to \partial\tilde{C}(R, M) \to \tilde{C}(R, M) \to C(R, M) \to 0$ gives the long exact sequence of cohomology:

$$(2.11.4) \qquad 0 \to H^0(\partial\tilde{C}(R, M)) \to \tilde{H}^0(R, M) \to H^0(R, M) \to$$

$$\to H^1(\partial\tilde{C}(R, M)) \to \tilde{H}^1(R, M) \to H^1(R, M) \to$$

$$\to H^2(\partial\tilde{C}(R, M)) \to \tilde{H}^2(R, M) \to H^2(R, M) \to \cdots$$

This cohomology theory describes extensions and deformations, just as any cohomology theory.

PROPOSITION 2.11. (a) $\tilde{H}^0(R, M) = \{m \in M \mid a_\lambda m = 0 \ \ \text{for all } a \in R\}$.

(b) *The isomorphism classes of extensions*

$$0 \to M \to E \to \mathbb{C} \to 0$$

of the trivial R-module $\mathbb{C}$ (∂ and R act by zero) by a conformal R-module M correspond bijectively to $H^0(R, M)$.

(c) *The isomorphism classes of $\mathbb{C}[\partial]$-split extensions*

$$0 \to M \to E \to N \to 0$$

of conformal modules over a conformal algebra R correspond bijectively to

$$H^1(R, \mathrm{Chom}(N, M)),$$

where M and N are assumed to be finite. If, in particular, $N = \mathbb{C}$ is the trivial module, then there exist no non-trivial $\mathbb{C}[\partial]$-split extensions.

(d) *Let C be a conformal R-module, considered as a conformal algebra with respect to the zero λ-bracket. Then the equivalence classes of $\mathbb{C}[\partial]$-split "abelian" extensions*

$$0 \to C \to \tilde{R} \to R \to 0$$

of the conformal algebra R correspond bijectively to $H^2(R, C)$.

(e) *The equivalence classes of first-order deformations of a conformal algebra R correspond bijectively to $H^2(R, R)$.*

PROOF. (a) The computation of $\tilde{H}^0(R, M)$ follows directly from the definition: for $m \in M = \tilde{C}^0(R, M)$ and $a \in R$, $(dm)_\lambda(a) = a_\lambda m$.

(b) Given an extension $0 \to M \to E \to \mathbb{C} \to 0$ of modules over a conformal algebra R, pick a splitting of the short exact sequence over $\mathbb{C}$, *i.e.*, assume that as a complex vector space, $E \simeq M \oplus \mathbb{C} = \{(m, n) \mid m \in M, n \in \mathbb{C}\}$. Define $f \in M$ by writing down the action of ∂ on the pair $(m, 1) \in E$:

$$(2.11.5) \qquad \partial(m, 1) = (\partial m + f, 0).$$

We claim that $f \in M = \tilde{C}^0(R, M)$ defines a 0-cocycle in the reduced complex $C(R, M)$ and thereby a class in $H^0(R, M)$.

To see that, define a 1-cochain $\gamma \in \tilde{C}^1(R, M)$ using the action of R on E:

$$(2.11.6) \qquad a_\lambda(m, 1) = (a_\lambda m + \gamma_\lambda(a), 0)$$

for $a \in R$. The property (2.11.1) of γ: $\gamma_\lambda(\partial a) = -\lambda\gamma_\lambda(a)$, follows from the fact that $(\partial a)_\lambda(m, 1) = -\lambda(a_\lambda(m, 1))$. The property $a_\lambda(\partial(m, 1)) = (\lambda + \partial)(a_\lambda(m, 1))$ of the action of R on E expands as

$$(2.11.7) \qquad (df)_\lambda = (\partial\gamma)_\lambda,$$

which means that $df = 0$ in the reduced complex.

If we choose another splitting $(m, n)'$ of the extension E, it will differ by an element $g \in M$:

$$(m, 1)' = (m + g, 1),$$

so that the new 0-cocycle becomes $f' = f + \partial g$, therefore defining the same cochain in the reduced complex.

If we have two isomorphic extensions and choose a compatible splitting over $\mathbb{C}$, we will get the same 0-cocycles corresponding to them. This proves that isomorphism classes of extensions give rise to elements of $H^0(R, M)$.

Conversely, given a cocycle in $C^0(R, M)$, we can choose a representative $f \in M$ of it to alter the natural $\mathbb{C}[\partial]$-module structure on $M \oplus \mathbb{C}$ by adding f to the action of ∂ on $M \oplus \mathbb{C}$ as in (2.11.5). This will obviously extend to an action of $\mathbb{C}[\partial]$. We can also alter the natural R-module structure by adding γ to the action of $a \in R$ as in (2.11.6), where γ is a solution of equation (2.11.7), which means that f is a cocycle in the reduced complex. This action satisfies $(M1)_\lambda$ because of (2.11.7) and the property (2.11.1) of γ, and it satisfies $(M2)_\lambda$ because $d\gamma = 0$, which follows from (2.11.7) and the fact that $\mathbb{C}[\partial]$ acts freely on basic 2-cochains.

By construction the natural map $M \to M \oplus \mathbb{C}$ and $M \oplus \mathbb{C} \to \mathbb{C}$ will be morphisms of $\mathbb{C}[\partial]$- and R-modules.

This construction of a new conformal module structure on $M \oplus \mathbb{C}$ involved a number of choices. The choice of a different representative $f' = f + \partial g$ defines an isomorphism of the two $\mathbb{C}[\partial]$-module structures on $M \oplus \mathbb{C}$, which automatically becomes an isomorphism of the corresponding R-module structures, because the corresponding γ's are unique. The 1-cochain γ is uniquely determined by f, because $\mathbb{C}[\partial]$ acts freely on the space $\tilde{C}^1(R, M)$ of basic 1-cochains.

(c) We will adjust the proof of (b) to the new situation. Given a $\mathbb{C}[\partial]$-split extension $0 \to M \to E \to N \to 0$ of modules over a conformal algebra R, pick a splitting of the short exact sequence over $\mathbb{C}[\partial]$, $i.e.$, assume that as a $\mathbb{C}[\partial]$-module, $E \simeq M \oplus N = \{(m, n) \mid m \in M, n \in N\}$. We are going to construct a reduced 1-cochain with coefficients in $\mathrm{Chom}(N, M)$ out of this data. Note that such cochains are linear maps $\gamma = \gamma_\lambda(a)_\mu$ from $R \otimes N$ to M depending on two variables λ and μ, considered modulo $\lambda - \mu$. Note that $\gamma_\lambda(a)_\mu \mod (\lambda - \mu)$ is fully determined by the restriction $\gamma_\lambda(a)_\lambda$ to the diagonal $\lambda = \mu$. Define a 1-cochain $\gamma \in C^1(R, \mathrm{Chom}(N, M))$ using the action of R on E:

$$(2.11.8) \qquad a_\lambda(m, n) = (a_\lambda m + \gamma_\lambda(a)_\lambda n, a_\lambda n)$$

for $a \in R$. The property (2.11.1) of γ: $\gamma_\lambda(\partial a)_\lambda = -\lambda \gamma_\lambda(a)_\lambda$, follows from the fact that $(\partial a)_\lambda(m, n) = -\lambda a_\lambda(m, n)$. The property $a_\lambda(\partial(m, n)) = (\lambda + \partial)(a_\lambda(m, n))$ of the action of R on E expands as

$$(2.11.9) \qquad\qquad (\partial\gamma)_\lambda = 0,$$

which means that $\gamma_\lambda(a)_\lambda$ is a conformal linear map $N \to M$. Finally, the module property $(M2)_\lambda$ for elements in E implies that $d\gamma = 0$.

If we choose another $\mathbb{C}[\partial]$-splitting $(m, n)'$ of the extension E, it will differ by an element $\beta \in \operatorname{Hom}_{\mathbb{C}[\partial]}(N, M)$:

$$(m, n)' = (m + \beta(n), n).$$

$\operatorname{Hom}_{\mathbb{C}[\partial]}(N, M)$ may be identified with the degree zero part of $\operatorname{Chom}(N, M)$, so that the new 1-cocycle becomes $\gamma' = \gamma + d\beta$, therefore defining the same cohomology class.

If we have two isomorphic extensions and choose a compatible splitting over $\mathbb{C}[\partial]$, we will have exactly the same 1-cocycles γ corresponding to them. This proves that isomorphism classes of extensions give rise to elements of $H^1(R, \operatorname{Chom}(N, M))$.

Conversely, given a cohomology class in $H^1(R, \operatorname{Chom}(N, M))$, we can choose a representative $\gamma \in C^1(R, \operatorname{Chom}(N, M))$ of it to alter the natural R-module structure on $M \oplus N$ by adding γ to the action of R on $M \oplus N$ as in (2.11.8). This action will satisfy $(M1)_\lambda$ because of (2.11.9) and (2.11.1). This action will define an R-module structure on $M \oplus N$, because $d\gamma = 0$ after the restriction to $\mu = \lambda_1 + \lambda_2$ in $\tilde{C}^2(R, \operatorname{Chom}(N, M))$.

By construction the natural mappings $M \to M \oplus N$ and $M \oplus N \to N$ will be morphisms of $\mathbb{C}[\partial]$- and R-modules.

This construction of a new conformal module structure on $M \oplus N$ is independent of the choice of a different representative $\gamma' = \gamma + d\beta$, because it defines an isomorphic structure of an R-module on $M \oplus N$.

Finally, if $N = \mathbb{C}$, then $\operatorname{Chom}(\mathbb{C}, M) = 0$, and therefore, there are no split extensions.

(d) Given a $\mathbb{C}[\partial]$-split extension of a conformal algebra R by a module C, choose a splitting $\tilde{R} = C \oplus R$. Then the bracket in $\tilde{R}$

$$[(0, a)_\lambda (0, b)] = (c_\lambda(a, b), a_\lambda b) \qquad \text{for } a, b \in R$$

defines a map $c\colon R \otimes R \to C[\lambda]$, satisfying $(C1)_\lambda$ and $(C1')_\lambda$ which we may combine with the natural mapping

$$C[\lambda] \to C[\lambda_1, \lambda_2]/(\partial + \lambda_1 + \lambda_2), \, p(\lambda) \mapsto p(\lambda_1),$$

to get the composite mapping, denoted c_{λ_1, λ_2}. It defines a 2-cochain, because it is obviously skew-symmetric and $(c_\lambda(\partial a, b), a_\lambda b) = [(0, \partial a)_\lambda(0, b)] = [\partial(0, a)_\lambda(0, b)] = -[\lambda(0, a)_\lambda(0, b)] = -\lambda(c_\lambda(a, b), a_\lambda b)$, which implies $c_{\lambda_1, \lambda_2}(\partial a, b) = -\lambda_1 c_{\lambda_1, \lambda_2}(a, b)$, and similarly, $c_{\lambda_1, \lambda_2}(a, \partial b) = -\lambda_2 c_{\lambda_1, \lambda_2}(a, b) \mod (\partial + \lambda_1 + \lambda_2)$. In fact, this 2-cochain c is a cocycle:

$$dc = a_{\lambda_1} c_{\lambda_2, \lambda_3}(b, c) - b_{\lambda_2} c_{\lambda_1, \lambda_3}(a, c) + c_{\lambda_3} c_{\lambda_1, \lambda_2}(a, b)$$

$$- c_{\lambda_1 + \lambda_2, \lambda_3}(a_{\lambda_1} b, c) + c_{\lambda_1 + \lambda_3, \lambda_2}(a_{\lambda_1} c, b) - c_{\lambda_2 + \lambda_3, \lambda_1}(b_{\lambda_2} c, a) = 0.$$

This is just because the Jacobi identity $(C3)_\lambda$, is satisfied in $\widetilde{R}$.

The construction of c assumed the choice of a splitting $\widetilde{R} = C \oplus R$. A different splitting would differ by a mapping $f\colon R \to C$, which can be thought of as $f\colon R \to C[\lambda]/(\partial + \lambda)$, which would contribute by df to c.

Thus, any extension determines a cohomology class in $H^2(R, C)$. The above arguments can be reversed to show that a class in the cohomology group defines an extension.

(e) Let $D = \mathbb{C}[t]/(t^2)$ be the algebra of *dual numbers*. Then a *first-order deformation* of a conformal algebra R is the structure of a conformal algebra over D on $R \otimes D$, so that the mapping $R \otimes D \to R$, $a \otimes p(t) \mapsto p(0)a$, is a morphism of conformal algebras. This means that classes of first-order deformations are in bijection with classes of $\mathbb{C}[\partial]$-split abelian extensions of R with the R-module R in the sense of part (d) of this theorem. Therefore, they are classified by $H^2(R, R)$. $\square$

Now I shall explain how the basic and reduced cohomology of a conformal algebra R with coefficients in a conformal R-module M is related to Lie algebra cohomology. Recall that M is canonically a module over the annihilation Lie algebra $\mathfrak{g}_- = (\mathrm{Lie}\, R)_-$ (see Remark 2.9a). Let $C(\mathfrak{g}_-, M) = \bigoplus_{n \in \mathbb{Z}_+} C^n(\mathfrak{g}_-, M)$ be the Chevalley–Eilenberg complex defining the cohomology of $\mathfrak{g}_-$ with coefficients in M. Recall that, by definition (see e.g. [**F**]), $C^n(\mathfrak{g}_-, M)$ is the space of skew-symmetric linear maps $\gamma \colon (\mathfrak{g}_-)^{\otimes n} \to M$ such that

$$\gamma(a_{1\,m_1} \otimes \cdots \otimes a_{n\,m_n}) = 0$$

for all but a finite number of $m_1, \ldots, m_n \in \mathbb{Z}_+$, where $a_1, \ldots, a_n \in R$, and $a_{i\,m_i} \in \mathfrak{g}_- = (\mathrm{Lie}\,R)_- = R[t]/(\partial + \partial_t)R[t]$ is the image of the element $a_i t^{m_i}$. Note that $C(\mathfrak{g}_-, M)$ has the following structure of a $\mathbb{C}[\partial]$-module:

$$(2.11.10) \quad (\partial\gamma)(a_1 \otimes \cdots \otimes a_n)$$

$$= \partial(\gamma(a_1 \otimes \cdots \otimes a_n)) - \sum_{i=1}^{n} \gamma(a_1 \otimes \cdots \otimes \partial a_i \otimes \cdots \otimes a_n).$$

THEOREM 2.11. *Let R be a conformal algebra, let $\mathfrak{g}_-$ denote its annihilation algebra and let M be a conformal R-module. Then*

(a) *There is a canonical isomorphism of complexes $\widetilde{C}(R, M)$ and $C(\mathfrak{g}_-, M)$, compatible with the action of $\mathbb{C}[\partial]$. Consequently, $\widetilde{H}(R, M) \simeq H(\mathfrak{g}_-, M)$ and the complex $C(R, M)$ is isomorphic to $C(\mathfrak{g}_-, M)/\partial C(\mathfrak{g}_-, M)$.*

(b) *Provided that M is a free $\mathbb{C}[\partial]$-module, the complex $C(R, M)$ is isomorphic to the subcomplex $C(\mathfrak{g}_-, V(M)_-)^\partial$ of ∂-invariant cochains in $C(\mathfrak{g}_-, V(M)_-)$.*

PROOF. (a) For a cochain $\gamma \in \widetilde{C}^n(R, M)$, we write

$$\gamma_{\lambda_1, \ldots, \lambda_n}(a_1, \ldots, a_n) = \sum_{m_1, \ldots, m_n \in \mathbb{Z}_+} \lambda_1^{(m_1)} \cdots \lambda_n^{(m_n)} \gamma_{(m_1, \ldots, m_n)}(a_1, \ldots, a_n).$$

In terms of the linear maps

$$\gamma_{(m_1, \ldots, m_n)} \colon R^{\otimes n} \to M, \qquad a_1 \otimes \cdots \otimes \gamma_n \mapsto \gamma_{(m_1, \ldots, m_n)}(a_1, \ldots, a_n),$$

the definition of $\widetilde{C}(R, M)$ translates as follows.

(i) for any $a_1, \ldots, a_n \in R$, $\gamma_{(m_1, \ldots, m_n)}(a_1, \ldots, a_n)$ is non-zero for only a finite number of $m_1, \ldots, m_n$,

(ii) $\gamma_{(m_1, \ldots, m_i, \ldots, m_n)}(a_1, \ldots, \partial a_i, \ldots, a_n)$

$= -m_i \gamma_{(m_1, \ldots, m_i-1, \ldots, m_n)}(a_1, \ldots, a_i, \ldots, a_n),$

(iii) γ is skew-symmetric with respect to simultaneous permutations of a_i's and m_i's.

The differential is given by:

$$(d\gamma)_{(m_1,\ldots,m_{n+1})}(a_1,\ldots,a_{n+1})$$

$$= \sum_{i=1}^{n+1}(-1)^{i+1}a_{i(m_i)}\gamma_{(m_1,\ldots,\widehat{m}_i,\ldots,m_{n+1})}(a_1,\ldots,\widehat{a}_i,\ldots,a_{n+1})$$

$$+ \sum_{\substack{i,j=1\\i<j}}^{n+1}\sum_{k=0}^{m_i}(-1)^{i+j}\binom{m_i}{k}\gamma_{(m_i+m_j-k,m_1,\ldots,\widehat{m}_i,\ldots,\widehat{m}_j,\ldots,m_{n+1})}(a_{i(k)}a_j,a_1,$$

$$\ldots,\widehat{a}_i,\ldots,\widehat{a}_j,\ldots,a_{n+1}).$$

Define linear maps $\phi^n\colon \widetilde{C}^n(R,M) \to C^n(\mathfrak{g}_-,M)$ by the formula

$$(\phi^n\gamma)(a_{1m_1}\otimes\cdots\otimes a_{nm_n}) = \gamma_{(m_1,\ldots,m_n)}(a_1,\ldots,a_n).$$

They are well-defined due to above condition (ii). Clearly, ϕ^n are bijective and, using (2.7.3), it is easy to see that $\phi^{n+1}\circ d = d\circ\phi^n$. Moreover, $\phi^n\circ\partial = \partial\circ\phi^n$, where ∂ acts on $\widetilde{C}(R,M)$ via (2.11.3) and on $C(\mathfrak{g}_-,M)$ via (2.11.10).

(b) Now we assume that M is a free $\mathbb{C}[\partial]$-module: $M = \mathbb{C}[\partial]\otimes_{\mathbb{C}} U$ for some vector space U. Then the $\mathfrak{g}_-$-module $V_- = V(M)_-$ is just $U[t]$ with

$$a_m(ut^n) = \sum_{j=0}^{m}\binom{m}{j}(a_{(j)}u)\,t^{m+n-j},\quad \partial(ut^n) = -nut^{n-1},$$

for $u\in U$, $a\in R$, see Section 2.9. In terms of the usual generating series $a_\lambda = \sum_{m\geq 0}\lambda^{(m)}a_m$, this can be rewritten as

$$a_\lambda(ut^n) = (a_\lambda u)\,t^n e^{t\lambda}.$$

Recall also that V_- is an R-module where $\mathbb{C}[\partial]$ acts by $\partial = -\partial_t$.

Let $\beta\in C^n(\mathfrak{g}_-,V_-)$. As in the proof of (a), consider the generating series

$$(2.11.11)\quad \beta_{\lambda_1,\ldots,\lambda_n;t}(a_1,\ldots,a_n)$$

$$= \sum_{m_1,\ldots,m_n\in\mathbb{Z}_+}\lambda_1^{(m_1)}\cdots\lambda_n^{(m_n)}\beta(a_{1m_1}\otimes\cdots\otimes a_{nm_n}).$$

By (2.11.10), ∂ acts on $\beta_{\lambda_1,\ldots,\lambda_n;t}$ as $-\partial_t + \sum\lambda_i$. Hence β is ∂-invariant iff

$$(2.11.12)\qquad \beta_{\lambda_1,\ldots,\lambda_n;t}(a_1,\ldots,a_n) = \gamma_{\lambda_1,\ldots,\lambda_n}(a_1,\ldots,a_n)\,e^{t\sum\lambda_i},$$

where $\gamma_{\lambda_1,\ldots,\lambda_n} = \beta_{\lambda_1,\ldots,\lambda_n;t}|_{t=0}$ takes values in U. Identifying U with $1 \otimes U \subset M$, we can consider γ as an element of $\widetilde{C}^n(R, M)$. It is easy to check that $\beta \mapsto \overline{\gamma} := \gamma$ mod $(\partial + \sum \lambda_i)$ is a chain map from $C(\mathfrak{g}_-, V_-)$ to $C(R, M)$.

Conversely, for $\overline{\gamma} \in C^n(R, M)$ choose a representative $\gamma \in \widetilde{C}^n(R, M)$ such that $\overline{\gamma} = \gamma$ mod $(\partial + \sum \lambda_i)$. Define $\beta \in C^n(\mathfrak{g}_-, V_-)^\partial$ by (2.11.11) and (2.11.12) with ∂ substituted by $-\partial_t$ in $\gamma_{\lambda_1,\ldots,\lambda_n}(a_1, \ldots, a_n) \in M = U[\partial]$. Then clearly, β is independent of the choice of γ.

The correspondence $\beta \leftrightarrow \overline{\gamma}$ establishes an isomorphism between $C(\mathfrak{g}_-, V_-)^\partial$ and $C(R, M)$. $\qquad\square$

EXAMPLE 2.11. Here we will compute the cohomology of the conformal algebra $\mathrm{Vir} = \mathbb{C}[\partial]L$, $[L_\lambda L] = (\partial + 2\lambda)L$ with trivial coefficients $M = \mathbb{C}$, where both ∂ and L act by zero. The answer is as follows:

$$\dim \widetilde{H}^q(\mathrm{Vir}, \mathbb{C}) = \begin{cases} 1 & \text{if } q = 0 \text{ or } 3, \\ 0 & \text{otherwise,} \end{cases} \qquad \dim H^q(\mathrm{Vir}, \mathbb{C}) = \begin{cases} 1 & \text{if } q = 0, 2, \text{ or } 3, \\ 0 & \text{otherwise.} \end{cases}$$

An n-cochain γ in this case is determined by its value on $L^{\otimes n}$:

$$P(\lambda_1, \ldots, \lambda_n) = \gamma_{\lambda_1,\ldots,\lambda_n}(L, \ldots, L).$$

Obviously, $P(\lambda_1, \ldots, \lambda_n)$ is a skew-symmetric polynomial with values in $\mathbb{C}$. The differential is then determined by the following formula:

$$(dP)(\lambda_1, \ldots, \lambda_{n+1}) = \sum_{\substack{i,j=1 \\ i<j}}^{n+1} (-1)^{i+j}(\lambda_i - \lambda_j)P(\lambda_i + \lambda_j, \lambda_1, \ldots, \widehat{\lambda_i}, \ldots, \widehat{\lambda_j}, \ldots, \lambda_{n+1}).$$

This describes the complex $\widetilde{C}(\mathrm{Vir}, \mathbb{C})$. The complex $C(\mathrm{Vir}, \mathbb{C})$ producing cohomology of Vir in degree n is nothing but the quotient of $\widetilde{C}^n(\mathrm{Vir}, \mathbb{C})$ by the ideal generated by $\sum_{i=1}^n \lambda_i$. In other words, $C^n(\mathrm{Vir}, \mathbb{C})$ is the space of regular (polynomial) functions on the hyperplane $\sum_{i=1}^n \lambda_i = 0$ in $\mathbb{C}^n$ which are skew-symmetric in the variables $\lambda_1, \ldots, \lambda_n$. (This complex appeared as an intermediate step in [GF1] of the calculation of cohomology of the Virasoro algebra, and its cohomology was computed there.) Consider the following homotopy operator $k \colon \widetilde{C}^q(\mathrm{Vir}, \mathbb{C}) \to \widetilde{C}^{q-1}(\mathrm{Vir}, \mathbb{C})$:

$$k(P) = (-1)^q \frac{\partial P}{\partial \lambda_q}\bigg|_{\lambda_q=0}.$$

A straightforward computation shows that $(dk + kd)P = (\deg P - q)P$ for $P \in \widetilde{C}^q(\mathrm{Vir}, \mathbb{C})$, where $\deg P$ is the total degree of P in $\lambda_1, \ldots, \lambda_q$. Thus, only those homogeneous cochains whose degree as a polynomial is equal to its degree as a cochain contribute to the cohomology of $\widetilde{C}(\mathrm{Vir}, \mathbb{C})$. These polynomials must be skew-symmetric and therefore divisible by $\Lambda_q = \prod_{i<j}(\lambda_i - \lambda_j)$, whose polynomial degree is $q(q-1)/2$. The quadratic inequality $q(q-1)/2 \leq q$ has $q = 0$, 1, 2, and 3 as the only integral solutions. For $q = 0$, the whole $\widetilde{C}^0 = \mathbb{C}$ contributes to $\widetilde{H}^0$. For $q = 1$, the only polynomial of degree 1 up to a constant factor is λ_1. Next, $d\lambda_1 = \lambda_2^2 - \lambda_1^2$, which is the only skew-symmetric polynomial of degree 2 in two variables. This shows that $\widetilde{H}^1 = \widetilde{H}^2 = 0$. Finally, for $q = 3$, the only skew-symmetric polynomial of degree 3 in 3 variables is Λ_3. It is easy to see that this polynomial represents a non-trivial class in the cohomology. Indeed, it is closed, because a skew-symmetric function in four variables has a degree at least 6, which is greater than $\deg(d\Lambda_3) = 4$, and Λ_3 is not a coboundary, because it can be the coboundary of a two-cochain of degree 2, which must be a constant multiple of $\lambda_2^2 - \lambda_1^2 = d\lambda_1$, whose coboundary is zero.

The computation of the cohomology of the quotient complex $C(\mathrm{Vir}, \mathbb{C})$ is based on the long exact sequence (2.11.4). By definition, $\partial \widetilde{C}^0 = 0$. To find the cohomology of $\partial \widetilde{C}(\mathrm{Vir}, \mathbb{C})$, define a homotopy $k_1 \colon \partial \widetilde{C}^q \to \partial \widetilde{C}^{q-1}$ as $k_1(\partial P) = \partial k(P)$, where $\partial = \sum_i \lambda_i$ and $P \in \widetilde{C}^q$. Then $(dk_1 + k_1 d)\partial P = (\deg P - q)\partial P$. As in the previous paragraph, this implies that $\deg P = q = 0$, 1, 2, or 3. Up to constant factors, the only polynomials in $\partial \widetilde{C}$ with this property are $P_1 = \lambda_1^2$ for $q = 1$, $P_2 = (\lambda_1 + \lambda_2)(\lambda_1^2 - \lambda_2^2)$ for $q = 2$, and $P_3 = (\lambda_1 + \lambda_2 + \lambda_3)\Lambda_3$ for $q = 3$. One computes: $dP_1 = -P_2$ and $dP_3 = 0$. Therefore $H^q(\partial \widetilde{C}) = 0$ for all q but $q = 3$, where it is 1-dimensional with the generator P_3. From the long exact sequence (2.11.4) we see that $H^0(\mathrm{Vir}, \mathbb{C}) = \mathbb{C}$ and $H^q(\mathrm{Vir}, \mathbb{C}) = 0$ for $q = 1, 4, 5, 6, \ldots$; $H^3(\mathrm{Vir}, \mathbb{C}) = \mathbb{C}\Lambda_3$ and $H^2(\mathrm{Vir}, \mathbb{C}) = \mathbb{C}(\lambda_1^3 - \lambda_2^3)$, because $d(\lambda_1^3 - \lambda_2^3) = P_3$.

In a similar fashion one can show that if ∂ acts on $\mathbb{C}$ non-trivially, then $\widetilde{H}^q(\mathrm{Vir}, \mathbb{C})$, remains the same, but $H^q(\mathrm{Vir}, \mathbb{C})$ becomes 0 for all q.

COROLLARY 2.11a. *The conformal algebra* Vir *has a non-trivial central extension by* $\mathbb{C}$ *iff* $\partial \mathbb{C} = 0$; *it is unique and is given by the 2-cocycle (2.7.19).*

For the calculation of basic and reduced cohomology of Cur $\mathfrak{g}$ with coefficients in $\mathbb{C}$ as well as of Vir and Cur $\mathfrak{g}$ with coefficients in $M(\Delta, \alpha)$ and $M(\tilde{U})$ respectively the reader is refered to [**BKV**]. One of the open problems is the calculation of cohomology of the conformal algebra gc_N. In order to demonstrate how beautiful the results are, let me state, in conclusion of this section, the answer for the Vir-modules $M(\Delta, \alpha)$:

(a) $H(\mathrm{Vir}, M_{(\Delta,\alpha)}) = 0$ if $\alpha \neq 0$ or if $\alpha = 0$ and $\Delta \neq 1 - (3r^2 + r)/2$ for any $r \in \mathbb{Z}$.

(b) $\dim H^q(\mathrm{Vir}, M_{(1-(3r^2\pm r)/2,0)}) = \begin{cases} 2 \text{ for } q = r + 1, \\ 1 \text{ for } q = r \text{ or } r + 2, \\ 0 \text{ otherwise.} \end{cases}$

Proof uses results of [**FeF**], [**F**] on cohomology of the Lie algebra of vector fields on $\mathbb{C}$ vanishing at 0 (see [**BKV**], Theorem 7.2 for details).

This theorem, along with Proposition 2.11d, implies the following corollary:

COROLLARY 2.11b. *There exists a non-trivial abelian extension* $0 \to M(\Delta, \alpha) \to R \to \mathrm{Vir} \to 0$ *iff* $\alpha = 0$ *and* $\Delta = 1, 0, -1, -4$ *or* -6.

This corollary (obtained earlier by M. Wakimoto and myself by a direct, but very lengthy, calculation) shows that a Levi splitting theorem does not hold in general. It is closely related to the calculations of [**R2**].

CHAPTER 3

Local fields

3.1. Normally ordered product

Fix a vector superspace $V = V_{\bar{0}} + V_{\bar{1}}$ (the space of states). Recall that a formal distribution

$$a(z) = \sum_{n \in \mathbf{Z}} a_{(n)} z^{-n-1}$$

with values in the ring $\mathrm{End}V$ (i.e., $a_{(n)} \in \mathrm{End}V$) is called a *field* if for any $v \in V$ one has:

$$a_{(n)}v = 0 \text{ for } n \gg 0.$$

This means that $a(z)v$ is a formal Laurent series in z (i.e., $a(z)v \in V[[z]][z^{-1}]$).

Note that in the expansion (cf. (2.2.5))

$$(3.1.1) \qquad [a(z), b(w)] = \sum_{j=0}^{\infty} c^j(w) \partial_w^{(j)} \delta(z - w) + b(z, w)^{+(z)}$$

all the coefficients $c^j(w)$ are fields provided that $b(w)$ is a field, due to formula (2.2.2):

$$(3.1.2) \qquad c^j(w) = \mathrm{Res}_z [a(z), b(w)](z - w)^j.$$

The *normally ordered product* of two fields $a(z)$ and $b(z)$ is defined by

$$(3.1.3) \qquad {:}\, a(z)b(z) \,{:} \,= a(z)_+ b(z) + (-1)^{p(a)p(b)} b(z)a(z)_-.$$

Since

$$(3.1.4) \qquad {:}\, a(z)b(z) \,{:}_{(n)} = \sum_{j=-1}^{-\infty} a_{(j)} b_{(n-j-1)} + (-1)^{p(a)p(b)} \sum_{j=0}^{\infty} b_{(n-j-1)} a_{(j)}$$

we see that when applied to $v \in V$ each of the two sums gives only a finite number of non-zero summands, hence $: a(z)b(z) :$ is a well defined formal distribution. Here we use that both $a(z)$ and $b(z)$ are fields; for general formal distribution one is able to define only the normally ordered product (2.3.5) in two indeterminates.

Moreover, it is clear from (3.1.3) that $: a(z)b(z) :$ is a field, since given $v \in V$, $b(z)v$ (resp. $a(z)_-v$) is a formal Laurent series (resp. a Laurent polynomial) in z, hence $a(z)_+b(z)v$ (resp. $b(z)a(z)_-v$) is a formal Laurent series in z.

Thus, the space of fields forms an algebra with respect to the normally ordered product (which is in general neither commutative nor associative).

Incidentally, it is straightforward to check that $: a(z)b(z) : -p(a,b) : b(z)a(z) :$ is a Lie superalgebra bracket (in spite of the non-associativity of the normally ordered product).[1]

The derivative $\partial a(z)$ of a field $a(z)$ is again a field and, thanks to (2.3.4), ∂ is a derivation of the normally ordered product:

$$(3.1.5) \qquad \partial : a(z)b(z) := : \partial a(z)b(z) : + : a(z)\partial b(z) : .$$

Due to the existence of the normally ordered product, one can define the n-th product between fields not only for n positive (see (2.3.8)), but also for n negative:

$$(3.1.6) \qquad a(z)_{(-n-1)}b(z) =: \partial^{(n)}a(z)b(z) :, \qquad n \in \mathbb{Z}_+.$$

It is tempting now, using these products and Taylor's formula (2.4.3), to rewrite the OPE (2.3.9a) of mutually local fields $a(z)$ and $b(z)$ in a "complete" form:

$$(3.1.7) \qquad a(z)b(w) = \sum_{j \in \mathbb{Z}} \frac{a(w)_{(j)}b(w)}{(z-w)^{j+1}}.$$

However, (3.1.7) makes no sense as an equality of formal distributions since different parts of it are expanded in different domains. (In the "graded" case one can give a meaning to (3.1.7) using analytic continuation.) Still, formula (3.1.7) can be used, up to an arbitrary order of $z - w$.

In order to state the result we need the notion of a *field in z and w*. This is a formal End$V$-valued distribution $a(z,w)$ such that $a(z,w)v \in V[[z,w]][z^{-1},w^{-1}]$.

For example, $: a(z)b(w) :$ is a field if $a(z)$ and $b(w)$ are fields. Note that a partial derivative of a field is a field and that $a(w,w)$ is a well defined field in the indeterminate w. The following is yet another version of Taylor's formula.

---

[1]This was pointed out to me by A. Radul.

PROPOSITION 3.1. *For any field $a(z,w)$ and any positive integer N there exist fields $c^j(w)$ $(0 \le j \le N-1)$ and a field $d^N(z,w)$ such that*

$$(3.1.8) \qquad a(z,w) = \sum_{j=0}^{N-1} c^j(w)(z-w)^j + (z-w)^N d^N(z,w).$$

The coefficients $c^j(w)$ are uniquely determined by this expansion and are given by the usual formula:

$$(3.1.9) \qquad c^j(w) = \partial_z^{(j)} a(z,w)\,|_{z=w}\,.$$

PROOF. The uniqueness of the $c^j(w)$ is proved in the usual way: differentiate j times (3.1.8) by z and let $z = w$. It suffices to prove existence of (3.1.8) for $N = 1$:

$$(3.1.10) \qquad a(z,w) - a(w,w) = (z-w)d(z,w) \text{ for some field } d(z,w),$$

since applying it again to $d(z,w)$ gives (3.1.8) for $N = 2$, etc. The proof of (3.1.10) is straightforward. $\qquad\square$

THEOREM 3.1. *Let $a(z)$ and $b(z)$ be mutually local fields and let N be a positive integer. Then there exists a field $d^N(z,w)$ such that in the domain $|z| > |w|$ one has:*

$$(3.1.11) \qquad a(z)b(w) = \sum_{j \ge -N} \frac{a(w)_{(j)} b(w)}{(z-w)^{j+1}} + (z-w)^N d^N(z,w).$$

The coefficients of $(z-w)^{-j-1}$ $(j \ge -N)$ in this expansion are uniquely determined.

PROOF. In view of (2.3.9a) and (3.1.5), the theorem is a consequence of Proposition 3.1 applied to the field $: a(z)b(w) :$. $\qquad\square$

Proposition 3.1 and Theorem 3.1 show that when calculating the OPE of local fields one can use Taylor's expansions up to the required order.

The following lemma will be used in the sequel.

LEMMA 3.1. *Let $a(z) = \sum_n a_{(n)} z^{-n-1}$ and $b(z) = \sum_n b_{(n)} z^{-n-1}$ be $EndV$-valued fields and let $|0\rangle \in V$ be a vector such that*

$$a_{(n)}|0\rangle = 0 \text{ and } b_{(n)}|0\rangle = 0 \text{ for } n \in \mathbb{Z}_+\,.$$

Then $(a(z)_{(n)} b(z))|0\rangle$ is a holomorphic V-valued formal distribution for all $n \in \mathbb{Z}$ with constant term $a_{(n)} b_{(-1)}|0\rangle$.

PROOF. Let $k \in \mathbb{Z}_+$ and consider separately two cases. The first case:

$$
\begin{aligned}
(a(z)_{(-k-1)}b(z))|0\rangle &= \; : \partial^{(k)}a(z)b(z) : |0\rangle = \partial^{(k)}(a(z))_+ b(z)|0\rangle \\
&= (\partial^{(k)}a(z))_+ b(z)_+ |0\rangle \, .
\end{aligned}
$$

We have used here (2.3.4). The second case:

$$
\begin{aligned}
(a(z)_{(k)}b(z))|0\rangle &= \sum_{j=0}^{k} \binom{k}{j}(-z)^{k-j}\left[a_{(j)}, b(z)\right]|0\rangle \\
&= \sum_{j=0}^{k} \binom{k}{j}(-z)^{k-j}a_{(j)}b(z)_+|0\rangle \, .
\end{aligned}
$$

Thus we see that in both cases lemma holds.

$\square$

It turns out that there is a nice unified formula for all the n-th products of fields ($n \in \mathbb{Z}$):

(3.1.12)
$$
a(w)_{(n)}b(w) = \mathrm{Res}_z \left(a(z)b(w)i_{z,w}(z-w)^n - (-1)^{p(a)p(b)}b(w)a(z)i_{w,z}(z-w)^n \right).
$$

Indeed, for $n \geq 0$ formula (3.1.12) obviously coincides with (2.3.8). For $n < 0$, (3.1.12) follows from the following formal Cauchy formulas for any formal distribution $a(z)$ and $k \in \mathbb{Z}_+$:

$$
\text{(3.1.13a)} \qquad \mathrm{Res}_z \, a(z)i_{z,w}\frac{1}{(z-w)^{k+1}} = \partial^{(k)}a(w)_+ \, ,
$$

$$
\text{(3.1.13b)} \qquad \mathrm{Res}_z \, a(z)i_{w,z}\frac{1}{(z-w)^{k+1}} = -\partial^{(k)}a(w)_- \, .
$$

It is immediate to check these formulas for $k = 0$; the general case follows by differentiating both sides by w k times.

3.2. Dong's lemma

Now we are in a position to prove the following important lemma (see [**Li**]).

LEMMA 3.2 (Dong). *If $a(z)$, $b(z)$ and $c(z)$ are pairwise mutually local fields (resp. formal distributions), then $a(z)_{(n)}b(z)$ and $c(z)$ are mutually local fields (resp. formal distributions) for all $n \in \mathbb{Z}$ (resp. $n \in \mathbb{Z}_+$). In particular $: a(z)b(z) :$ and $c(z)$ are mutually local fields provided that $a(z)$, $b(z)$ and $c(z)$ are.*

PROOF. It suffices to show that for $M \gg 0$:

$$(3.2.1) \qquad (z_2 - z_3)^M A = (z_2 - z_3)^M B,$$

where

$$
(3.2.2a) \quad A = i_{z_1,z_2}(z_1 - z_2)^n a(z_1)b(z_2)c(z_3) \\
- (-1)^{p(a)p(b)} i_{z_2,z_1}(z_1 - z_2)^n b(z_2)a(z_1)c(z_3),
$$

$$
(3.2.2b) \quad B = (-1)^{p(c)(p(a)+p(b))} \left(i_{z_1,z_2}(z_1 - z_2)^n c(z_3)a(z_1)b(z_2) \right. \\
\left. - (-1)^{p(a)p(b)} i_{z_2,z_1}(z_1 - z_2)^n c(z_3)b(z_2)a(z_1) \right).
$$

Indeed, taking Res_{z_1} of both sides of (3.2.1) and letting $z_2 = z$, $z_3 = w$ gives the result due to (3.1.12).

The pairwise locality means that for $r \gg 0$:

$$
(3.2.3a) \quad (z_1 - z_2)^r a(z_1)b(z_2) = (z_1 - z_2)^r (-1)^{p(a)p(b)} b(z_2)a(z_1),
$$

$$
(3.2.3b) \quad (z_2 - z_3)^r b(z_2)c(z_3) = (z_2 - z_3)^r (-1)^{p(b)p(c)} c(z_3)b(z_2),
$$

$$
(3.2.3c) \quad (z_1 - z_3)^r a(z_1)c(z_3) = (z_1 - z_3)^r (-1)^{p(a)p(c)} c(z_3)a(z_1).
$$

Taking r sufficiently large, we may assume that $n \geq -r$. Take $M = 4r$ and use

$$(z_2 - z_3)^{3r} = \sum_{s=0}^{3r} \binom{3r}{s} (z_2 - z_1)^{3r-s}(z_1 - z_3)^s.$$

Then the left-hand side of (3.2.1) becomes

$$(3.2.4) \qquad \sum_{s=0}^{3r} \binom{3r}{s} (z_2 - z_1)^{3r-s}(z_1 - z_3)^s (z_2 - z_3)^r A.$$

If $3r - s + n \geq r$, then $(z_1 - z_2)^{3r-s} i_{z_1,z_2}(z_1 - z_2)^n = (z_1 - z_2)^{r'}$ where $r' \geq r$, hence due to (3.2.3a) the s-th summand in (3.2.4) is 0 for $0 \leq s \leq r$. Hence the left-hand side of (3.2.1) equals

$$(3.2.5a) \qquad \sum_{s=r+1}^{3r} \binom{3r}{s} (z_2 - z_1)^{3r-s}(z_1 - z_3)^s (z_2 - z_3)^r A.$$

Similarly the right-hand side of (3.2.1) equals

$$(3.2.5b) \qquad \sum_{s=r+1}^{3r} \binom{3r}{s} (z_2 - z_1)^{3r-s}(z_1 - z_3)^s (z_2 - z_3)^r B.$$

Due to (3.2.3b and c), (3.2.5a) is equal to (3.2.5b). $\qquad\square$

Let $g\ell f(V)$ denote the space (over $\mathbb{C}$) of all fields with values in $\mathrm{End}V$. As we have seen, $g\ell f(V)$ is closed under all the products $a(z)_{(n)}b(z)$, $n \in \mathbb{Z}$. This is called the *general linear field algebra*.

DEFINITION 3.2. A subspace F of $g\ell f(V)$ containing the identity operator I_V and closed under all the products $a(z)_{(n)}b(z)$ (then automatically $\partial_z F \subset F$) is called a *linear field algebra*.[2] A linear field algebra is called *local* if it consists of mutually local fields.

REMARK 3.2. A subspace F of $g\ell f(V)$ is a linear field algebra iff $I_V \in F$, $\partial F \subset F$, F is closed under normally ordered product and F is closed under OPE (i.e., all the OPE coefficients given by (3.1.2) are in F).

One says that a collection of fields *generates* a field algebra F if F is the minimal field algebra containing these fields. Dong's lemma implies

COROLLARY 3.2. *A linear field algebra generated by a collection of mutually local fields is local.*

Let $F \subset gl f(V)$ be a linear field algebra. Then we may associate to any $a \in F$ a formal distribution with values in $\mathrm{End}_{\mathbb{C}}F$:

$$Y(a, x) = \sum_{n \in \mathbb{Z}} x^{-n-1} a_{(n)} \,.$$

Explicitly, using (3.1.12), this formal distribution can be written as follows:

(3.2.6) $$Y(a(w), x)b(w)$$

$$= \mathrm{Res}_z (a(z)b(w)i_{z,w}\delta((z-w)-x) - p(a,b)b(w)a(z)i_{w,z}\delta((z-w)-x))\,.$$

This formal distribution is a field if F is a local field algebra.

The following proposition will be used in the sequel.

PROPOSITION 3.2. *If $a(z), b(z)$ are elements of a linear field algebra F and $N > 0$ is such that $(z-w)^N [a(z), b(w)] = 0$, then*

(3.2.7) $$(x-y)^N [Y(a(w), x), Y(b(w), y)] = 0\,.$$

---

[2]Lian and Zuckerman [**LZ**] use the term "quantum operator algebra."

PROOF. It is straightforward to see from (3.2.6) that

$$[Y(a(w), x), Y(b(w), y)]c(w)$$

$$= \mathrm{Res}_{z_1} \mathrm{Res}_{z_2}([a(z_1), b(z_2)]c(w)i_{z_1,w}i_{z_2,w}$$

$$- p(a,c)p(b,c)c(w)[a(z_1), b(z_2)]i_{w,z_1}i_{w,z_2})$$

$$\times \delta((z_1 - w) - x)\delta((z_2 - w) - y).$$

Since $x - y = ((z_2 - w) - y) - ((z_1 - w) - x) + (z_1 - z_2)$, (3.2.7) follows using Proposition 2.1e (for $j = 0$). $\qquad\square$

3.3. Wick's theorem and a "non-commutative" generalization

The normally ordered product of more than two fields $a^1(z), a^2(z), \ldots, a^N(z)$ is defined inductively "from right to left":

$$(3.3.1) \qquad : a^1(z)a^2(z)\cdots a^N(z) := : a^1(z)\cdots : a^{N-1}(z)a^N(z) : \cdots : .$$

This is a sum of 2^N terms of the form

$$(3.3.2) \qquad \pm a^{i_1}(z)_+ a^{i_2}(z)_+ \cdots a^{j_1}(z)_- a^{j_2}(z)_- \cdots ,$$

where $i_1 < i_2 < \cdots, j_1 > j_2 > \cdots$ is a permutation of the index set $\{1, \ldots, N\}$ and $\pm$ is the sign of this permutation from which the indices of even fields are removed.

REMARK 3.3. It is clear from (3.3.2) that if $\left[a^i(z)_\pm, a^j(z)_\pm\right] = 0$ for all i and j, then $: a^1(z)\cdots a^N(z) : \; = \pm \; : a^{i_1}(z)\cdots a^{i_N}(z) :$ where $\pm$ is the sign of the permutation of $i_1, \cdots, i_N$ from which the indices of even fields are removed. It follows that in this case the normally ordered product is (super)commutative. However, it is not associative, as Example 4.8 in Section 4.8 shows.

The following well-known simple theorem is extremely useful for calculating the OPE of two normally ordered products of "free" fields.

THEOREM 3.3 (Wick theorem). *Let $a^1(z), \ldots, a^M(z)$ and $b^1(z), \ldots, b^N(z)$ be two collections of fields such that the following properties hold:*

(i) $\left[[a^i(z)_-, b^j(w)], c^k(z)_\pm\right] = 0$ *for all i, j, k, and $c = a$ or b,*

(ii) $\left[a^i(z)_\pm, b^j(w)_\pm\right] = 0$ *for all i and j.*

Let $[a^i b^j] = [a^i(z)_-, b^j(w)]$ denote the "contraction" of $a^i(z)$ and $b^j(w)$. Then one has:

$$(3.3.3) \quad : a^1(z) \cdots a^M(z) :: b^1(w) \cdots b^N(w) := \sum_{s=0}^{\min(M,N)} \sum_{\substack{i_1 < \cdots < i_s \\ j_1 \neq \cdots \neq j_s}}$$

$$\left(\pm \left[a^{i_1} b^{j_1} \right] \cdots \left[a^{i_s} b^{j_s} \right] : a^1(z) \cdots a^M(z) b^1(w) \cdots b^N(w) :_{(i_1,\ldots,i_s;j_1,\ldots,j_s)} \right)$$

where the subscript $(i_1 \cdots i_s; j_1 \cdots j_s)$ means that the fields $a^{i_1}(z)$, ..., $a^{i_s}(z)$, $b^{j_1}(w)$, ..., $b^{j_s}(w)$ are removed, and the sign $\pm$ is obtained by the usual super rule: each permutation of the adjacent odd fields changes the sign.

PROOF. The typical term on the left-hand side of (3.3.3) is

$$\left(\pm a^{j_1}(z)_+ a^{j_2}(z)_+ \cdots a^{i_1}(z)_- a^{i_2}(z)_- \cdots \right) \left(\pm b^{j_1'}(w)_+ b^{j_2'}(w)_+ \cdots b^{i_1'}(w)_- b^{i_2'}(w)_- \cdots \right)$$

and we have to move the $a^i(z)_-$ across the $b^j(w)_+$ in order to bring this product to the normally ordered form (3.3.2). But due to the condition (ii) of the theorem,

$$(3.3.4) \qquad a^i(z)_- b^j(w)_+ = (-1)^{p(a^i)p(b^j)} b^j(w)_+ a^i(z)_- + \left[a^i(z)_-, b^j(w) \right] .$$

Due to condition (i) the contractions commute with all fields, hence can be moved to the left. This proves (3.3.3). $\qquad\qquad \square$

DEFINITION 3.3. A collection of fields $\{a^\alpha(z)\}$ is called a *free field theory* if all of these fields are mutually local and all the coefficients of the singular parts of the OPE are multiples of the identity.

By Remark 3.3, normally ordered products of free fields are, up to the sign, independent of the order. The OPE between these normally ordered products can be calculated using Wick's formula (3.3.3) and Taylor's formula (3.1.8).

Now we turn to a generalization of Wick's formula for arbitrary fields. First, we prove an analogue of Proposition 2.3 for all n-th products of fields.

PROPOSITION 3.3. (a) *For any two fields $a(w)$ and $b(w)$ and any $n \in \mathbb{Z}$ one has:*

$$(3.3.5a) \qquad\qquad \partial a(w)_{(n)} b(w) = -n a(w)_{(n-1)} b(w) ,$$

$$(3.3.5b) \qquad a(w)_{(n)} \partial b(w) = \partial_w (a(w)_{(n)} b(w)) + n a(w)_{(n-1)} b(w) .$$

Hence, ∂ is a derivation of all n-th products.

(b) *For any mutually local fields $a(w)$ and $b(w)$, and for any $n \in \mathbb{Z}$ one has:*

$$(3.3.6) \qquad a(w)_{(n)} b(w) = -p(a,b) \sum_{j=0}^{\infty} (-1)^{j+n} \partial_w^{(j)} \left(b(w)_{(n+j)} a(w) \right).$$

(c) *For any three fields $a(w)$, $b(w)$, and $c(w)$ and for any $m \in \mathbb{Z}_+$, $n \in \mathbb{Z}$ one has:*

$$(3.3.7) \qquad \begin{aligned} a(w)_{(m)} \left(b(w)_{(n)} c(w) \right) &= \sum_{j=0}^{m} \binom{m}{j} \left(a(w)_{(j)} b(w) \right)_{(m+n-j)} c(w) \\ &\quad + p(a,b) b(w)_{(n)} \left(a(w)_{(m)} c(w) \right). \end{aligned}$$

PROOF. The proof of (a) is straightforward.

We have by (3.1.11) in the domain $|z| > |w|$:

$$(3.3.8) \qquad b(z)a(w) = \sum_{k \geq -N} \frac{b(w)_{(k)} a(w)}{(z-w)^{k+1}} + (z-w)^N d(z,w).$$

Using locality (see Theorem 2.3(iii)) and exchanging z and w we obtain from (3.3.8) in the domain $|z| > |w|$:

$$p(a,b) a(z) b(w) = \sum_{n \geq -N} \frac{b(z)_{(n)} a(z)}{(w-z)^{n+1}} + (w-z)^N d(w,z).$$

Applying Proposition 3.1 to $a(z,w) = b(z)_{(n)} a(z)$ we rewrite this as:

(3.3.9)

$$p(a,b) a(z) b(w) = \sum_{n \geq -N} (-1)^{n+1} \sum_{j \geq 0} \frac{\partial^{(j)} (b(w)_{(n)} a(w))}{(z-w)^{n+1-j}} + (z-w)^N d_1(w,z).$$

Comparing the coefficients of $(z-w)^{-k-1}$ in (3.3.8) (where a and b are exchanged) and in (3.3.9) we get (b).

An equivalent form of (3.3.7) is the following formula:

$$(3.3.10) \qquad \begin{aligned} [a(w)_\lambda (b(w)_{(n)} c(w))] &= p(a,b) b(w)_{(n)} [a(w)_\lambda c(w)] \\ &\quad + \sum_{k=0}^{\infty} \lambda^{(k)} [a(w)_\lambda b(w)]_{(n+k)} c(w) \end{aligned}$$

(where $[a(w)_\lambda b(w)]$ is defined by (2.3.11)). The proof of (3.3.10) is straightforward using the identity

$$[a, bc] = [a,b]c + p(a,b) b[a,c].$$

Indeed, the left-hand side of (3.3.10) is

$$\mathrm{Res}_z \, e^{\lambda(z-w)} [a(z), b(w)_{(n)} c(w)] = A - p(b,c) B,$$

where

$$A \;=\; \mathrm{Res}_z\,\mathrm{Res}_u\,e^{\lambda(z-w)}[a(z),b(u)c(w)]i_{u,w}(u-w)^n\,,$$

$$B \;=\; \mathrm{Res}_z\,\mathrm{Res}_u\,e^{\lambda(z-w)}[a(z),c(w)b(u)]i_{w,u}(u-w)^n\,.$$

We have:

$$
\begin{aligned}
A \;&=\; \mathrm{Res}_z\,\mathrm{Res}_u(e^{\lambda(z-w)}[a(z),b(u)]c(w)i_{u,w}(u-w)^n\\[4pt]
&+\; p(a,b)b(u)[a(z),c(w)]i_{u,w}(u-w)^n)\\[4pt]
&=\; \mathrm{Res}_z\,\mathrm{Res}_u(e^{\lambda(z-u)}e^{\lambda(u-w)}[a(z),b(u)]c(w)i_{u,w}(u-w)^n\\[4pt]
&+\; p(a,b)\,\mathrm{Res}_u\,b(u)[a(w)_\lambda c(w)]i_{u,w}(u-w)^n)\\[4pt]
&=\; \mathrm{Res}_u([a(u)_\lambda b(u)]e^{\lambda(u-w)}c(w)\\[4pt]
&+\; p(a,b)b(u)[a(w)_\lambda c(w)])i_{u,w}(u-w)^n\,.
\end{aligned}
$$

Similarly we obtain:

$$B = \mathrm{Res}_u([a(w)_\lambda c(w)]b(u) + p(a,c)c(w)e^{\lambda(u-w)}[a(u)_\lambda b(u)])i_{w,u}(u-w)^n\,.$$

These two equations give (3.3.10). $\square$

The special case of (3.3.7) for $n = -1$ is called the "non-commutative" Wick formula ($m \in \mathbb{Z}_+$):

$$(3.3.11)\quad a(z)_{(m)}:b(z)c(z): =: \big(a(z)_{(m)}b(z)\big)\,c(z):$$

$$+\,p(a,b):b(z)\big(a(z)_{(m)}c(z)\big): + \sum_{j=0}^{m-1}\binom{m}{j}\big(a(z)_{(j)}b(z)\big)_{(m-1-j)}c(z).$$

Note that for free fields the "correcting" sum in (3.3.11) vanishes and we recover the usual Wick formula.

Formulas (3.3.6) and (3.3.11) allow one to calculate OPE of arbitrary normally ordered products of pairwise local fields knowing the OPE of these fields if they form a closed system under n-th products for $n \in \mathbb{Z}_+$. In fact there is a Mathematica package [T] which provides a computer program for these calculations. The earliest known to me reference where formula (3.3.11) is explicitly written down and systematically used is the paper [BBSS].

In the case of $n = -1$ formula (3.3.10) can be written in the following beautiful form (equivalent to (3.3.11)):

$$(3.3.12) \qquad [a(w)_\lambda : b(w)c(w) :]$$

$$=: [a(w)_\lambda b(w)]c(w) : +p(a,b) : b(w)[a(w)_\lambda c(w)] :$$

$$+ \int_0^\lambda [[a(w)_\lambda b(w)]_\mu c(w)]\, d\mu .$$

3.4. Bounded and field representations of formal distribution Lie superalgebras

DEFINITION 3.4a. Let $\mathfrak{g}$ be a formal distribution Lie superalgebra, i.e. a Lie superalgebra spanned by coefficients of a family of mutually local formal distributions $\{a^\alpha(z)\}_{\alpha \in A}$ (A an index set). A representation of $\mathfrak{g}$ in a vector space V is called a *field representation* if all the $a^\alpha(z)$ are represented by fields, i.e. for each $v \in V$ and $\alpha \in A$ one has

$$a^\alpha_{(n)} v = 0 \text{ for } n \gg 0 .$$

An important problem of quantum field theory is the construction of local linear field algebras. The usual way of doing this is to take a field representation of a formal distribution Lie superalgebra; then the fields representing the $a^\alpha(z)$ generate a local linear field algebra.

Field representations are usually constructed by means of induced modules. Recall that for a Lie superalgebra $\mathfrak{g}$ and a representation π of its subalgebra $\mathfrak{p}$ in a vector space W the *induced* $\mathfrak{g}$-module is the vector space

$$\text{Ind}_\mathfrak{p}^\mathfrak{g} \pi \quad := \quad U(\mathfrak{g}) \otimes_{U(\mathfrak{p})} W$$

$$\equiv \quad (U(\mathfrak{g}) \otimes W) / U(\mathfrak{g}) \langle p \otimes w - 1 \otimes \pi(p)w \,|\, p \in \mathfrak{p}, w \in W \rangle$$

on which $g \in \mathfrak{g}$ acts by left multiplication on the 1st factor.

Let $\mathfrak{g}$ be a Lie superalgebra spanned by coefficients of mutually local formal distributions $\{a^\alpha(z)\}_{\alpha \in A}$ and assume that the $\mathbb{C}[\partial]$-span of the $a^\alpha(z)$ is closed under all n-th products, $n \in \mathbb{Z}_+$ (cf. Corollary 4.7). Let

$$(3.4.1) \qquad\qquad \mathfrak{g}_- = \mathbb{C}\text{-span of } \left\{ a^\alpha_{(n)} | \alpha \in A, n \in \mathbb{Z}_+ \right\} .$$

Due to Theorem 2.3(iv), $\mathfrak{g}_-$ is a subalgebra of $\mathfrak{g}$. It is called the *annihilation subalgebra* (cf. Section (2.9)). Let π be a representation of $\mathfrak{g}_-$ in a vector space W such that for any $w \in W$:

$$\pi\left(a^\alpha_{(n)}\right) w = 0 \quad \text{for} \quad n \gg 0.$$

Then the induced $\mathfrak{g}$-module $\mathrm{Ind}^{\mathfrak{g}}_{\mathfrak{g}_-}\pi$ is a field representation. Indeed, one proves by induction on k (using Theorem 2.3(iv)) that

$$a^\alpha_{(n)}\left(a^{\alpha_1}_{(n_1)}\cdots a^{\alpha_k}_{(n_k)}w\right) = 0 \quad \text{for} \quad n \gg 0.$$

Unfortunately, even the oscillator algebra has a lot of pathological irreducible field representations. The additional requirement of "boundedness" removes these pathologies.

We shall now assume that the formal distribution Lie superalgebra $\mathfrak{g}$ is *graded*. This means that we have a diagonalizable derivation H of the Lie superalgebra $\mathfrak{g}$ such that for some $\Delta_\alpha \in \mathbb{R}$:

$$(3.4.2) \qquad\qquad Ha^\alpha(z) = \left(z\partial_z + \Delta_\alpha\right)a^\alpha(z)$$

i.e., $a^\alpha(z)$ is an eigendistribution for H of conformal weight Δ_α. Writing $a^\alpha(z) = \sum_{n\in-\Delta_\alpha+\mathbb{Z}} a^\alpha_n z^{-n-\Delta_\alpha}$ we have, due to (2.6.1):

$$Ha^\alpha_n = -na^\alpha_n.$$

Hence $\mathfrak{g}$ is a $\mathbb{R}$-graded Lie superalgebra by eigenspaces of H:

$$(3.4.3) \qquad\qquad \mathfrak{g} = \oplus_n \mathfrak{g}_n, \quad [\mathfrak{g}_m, \mathfrak{g}_n] \subset \mathfrak{g}_{m+n}.$$

Let

$$\mathfrak{g}^{\geq} = \oplus_{n\geq 0}\mathfrak{g}_n, \quad \mathfrak{g}^{>0} = \oplus_{n>0}\mathfrak{g}_n, \ \mathfrak{g}^{<0} = \oplus_{n<0}\mathfrak{g}_n.$$

We have the *triangular decomposition*:

$$\mathfrak{g} = \mathfrak{g}^< + \mathfrak{g}_0 + \mathfrak{g}^>.$$

DEFINITION 3.4b. A representation in a vector space V of graded formal distribution Lie superalgebra $\mathfrak{g}$ is called *bounded*[3] if the subalgebra $\mathfrak{g}^{>0}$ acts locally

---

[3]This terminology differs from that of [**K2**], where field modules are called "restricted" and bounded modules are more or less the "category $\mathcal{O}$" modules.

nilpotently on V, i.e., for any $v \in V$ there exists $n > 0$ such that $g_1 \cdots g_n v = 0$ for any n elements $g_1, \ldots, g_n$ of $\mathfrak{g}^{>0}$.

Recall that a $\mathfrak{g}$-module V is called *graded* if $V = \oplus_{j \in \mathbb{R}} V_j$ and $\mathfrak{g}_m V_n \subset V_{m+n}$.

Consider a representation π of the subalgebra $\mathfrak{g}_0$, extend it to $\mathfrak{g}^{\geq}$ by letting $\pi\left(\mathfrak{g}^{>0}\right) = 0$, and let

$$\tilde{V}(\pi) = \operatorname{Ind}_{\mathfrak{g}^{\geq}}^{\mathfrak{g}} \pi.$$

The $\mathfrak{g}$-module $\tilde{V}(\pi)$ is called the (generalized) *Verma module* associated to π. Note that this is a graded module, the gradation being induced by $\mathbb{R}$-gradation (3.4.3):

$$(3.4.4) \qquad \tilde{V}(\pi) = \bigoplus_{n \geq 0} \tilde{V}(\pi)_n,$$

so that the representation of $\mathfrak{g}_0$ in $\tilde{V}(\pi)_0$ is π. It follows from (3.4.4) that the representation of $\mathfrak{g}$ in $\tilde{V}(\pi)$ is a bounded field representation.

Denote by $J(\pi)$ the sum of all $\mathfrak{g}$-submodules contained in $\bigoplus_{n>0} \tilde{V}(\pi)_n$, and let

$$V(\pi) = \tilde{V}(\pi)/J(\pi).$$

It is clear that $J(\pi)$ is a graded submodule, hence $V(\pi)$ is a graded module.

A vector v of a $\mathfrak{g}$-module V is called *singular* if $\mathfrak{g}^{>0}v = 0$.

The proof of the following proposition is straightforward.

PROPOSITION 3.4. (a) *A graded bounded $\mathfrak{g}$-module $V = \bigoplus_j V_j$ is irreducible iff all its singular vectors have minimal grade d and the representation of $\mathfrak{g}_0$ in V_d is irreducible.*

(b) *The map $\pi \mapsto V(\pi)$ gives us a bijection between the set of all (up to isomorphism) irreducible $\mathfrak{g}_0$-modules and the set of all (up to isomorphism and shift of grade) irreducible bounded $\mathfrak{g}$-modules.* $\qquad\square$

3.5. Free (super)bosons

Let $\mathfrak{h}$ be a finite-dimensional superspace with a non-degenerate supersymmetric bilinear form $(.|.)$. Viewing $\mathfrak{h}$ as a commutative Lie superalgebra, we may consider its affinization (see Section 2.5):

$$\hat{\mathfrak{h}} = \mathfrak{h}[t, t^{-1}] + \mathbb{C}K$$

with commutation relations ($m, n \in \mathbb{Z}$; $a, b \in \mathfrak{h}$):

$$[a_m, b_n] = m(a|b)\delta_{m,-n}K, \quad [K, \hat{\mathfrak{h}}] = 0, \tag{3.5.1}$$

where a_m stands for at^m. Then the currents

$$a(z) = \sum_{n \in \mathbb{Z}} a_n z^{-n-1}, \quad a \in \mathfrak{h},$$

are mutually local with the OPE (cf. (2.5.6)):

$$a(z)b(w) \sim \frac{(a|b)K}{(z-w)^2}. \tag{3.5.2}$$

It is natural to call $\hat{\mathfrak{h}}$ the Weyl affinization of $\mathfrak{h}$ (vs. the Clifford affinization Cl_A discussed in Section 2.5 and in the next section). The different nature of notation stems from the difference of the generalizations of these two affinizations to the non-commutative case discussed in Section 2.5.

Consider a field representation of the Lie superalgebra $\hat{\mathfrak{h}}$ in a vector space V. Then we get a set of mutually local fields with the OPE (3.5.2), called a system of *free bosons* (sometimes called free superbosons if $\mathfrak{h}_{\bar{1}} \neq 0$). Note that these fields satisfy the conditions of Wick's theorem.

Choose bases $\{a^i\}$ and $\{b^i\}$ of $\mathfrak{h}$ consistent with the $\mathbb{Z}_2$-gradation such that

$$(b^i|a^j) = \delta_{ij}. \tag{3.5.3}$$

Such bases are called *dual*. Then for any $h \in \mathfrak{h}$ we have:

$$h = \sum_i (b^i|h)a^i = \sum_i (h|a^i)b^i. \tag{3.5.4}$$

Consider now the field

$$S(z) = \frac{1}{2}\sum_i : a^i(z)b^i(z) : . \tag{3.5.5}$$

Using Wick's theorem, calculate the following OPE:

$$S(z)a(w) \sim \frac{1}{2}\sum_i \frac{(b^i|a)}{(z-w)^2}a^i(z)K + \frac{1}{2}\sum_i (-1)^{p(b^i)p(a)}\frac{(a^i|a)}{(z-w)^2}b^i(z)K .$$

Using (3.5.4), we obtain ($a \in \mathfrak{h}$):

$$S(z)a(w) \sim \frac{a(z)}{(z-w)^2}K \sim \left(\frac{a(w)}{(z-w)^2} + \frac{\partial a(w)}{z-w}\right)K . \tag{3.5.6}$$

In the last part of (3.5.6) we used Taylor's formula.

Suppose now that $K = kI_V$ where the *affine central charge* k is a non-zero number. Let

$$(3.5.7) \qquad L(z) = \frac{1}{k} S(z).$$

Then (3.5.6) gives us ($a \in \mathfrak{h}$):

$$(3.5.8) \qquad L(z)a(w) \sim \frac{a(z)}{(z-w)^2} \sim \frac{a(w)}{(z-w)^2} + \frac{\partial a(w)}{z-w}.$$

Writing $L(z) = \sum_{n \in \mathbb{Z}} L_n z^{-n-2}$, we obtain, due to Table OPE (Sec. 2.6):

$$(3.5.9) \qquad [L_m, a_n] = -n a_{m+n}, \quad m, n \in \mathbb{Z}.$$

Noting that

$$L_0 = \frac{1}{2k} \sum_i a_0^i b_0^i + H,$$

where

$$H = \frac{1}{2k} \sum_i \sum_{n>0} \left(a_{-n}^i b_n^i + (-1)^{p(a^i)} b_{-n}^i a_n^i \right)$$

and that the elements a_0 lie in the center of $\hat{\mathfrak{h}}$, we see from (3.5.9), in particular, that

$$(3.5.10) \qquad [H, a_n] = -n a_n.$$

In other words, $\mathrm{ad}H$ is a Hamiltonian and all fields $a(z)$ have conformal weight 1. (Of course, it is even easier to check (3.5.9) and (3.5.10) directly.)

Note that (3.5.9) for $m = -1$ and $m = 0$ means

$$[L_{-1}, a(z)] = \partial a(z), \quad [L_0, a(z)] = (z\partial + 1)a(z).$$

It follows easily that $L(z)$ satisfies (2.6.4). Since also $L(z)$ is a local field whose OPE with itself, by Wick's theorem, has the form (2.6.3) we obtain by Theorem 2.6b that $L(z)$ is a Virasoro field. (Of course, it is easy to see this directly using Wick's theorem.) In order to compute the central charge, we need to compute the $s = 2$ term of $L(z)L(w)$ in Wick's formula (3.3.3), which is $\frac{1}{2} \operatorname{sdim} \mathfrak{h}/(z-w)^4$. Thus we obtain

$$(3.5.11) \qquad \text{central charge of } L(z) = \operatorname{sdim} \mathfrak{h}.$$

Since $\partial a(z)$ has conformal weight 2 we can construct the following family of local fields of conformal weight 2:

$$L^b(z) = L(z) + \partial b(z), \quad b \in \mathfrak{h}_{\bar{0}}.$$

As usual, we let $L^b(z) = \sum_n L^b_n z^{-n-2}$. It follows from (3.5.2) and (3.5.8) that

$$(3.5.12) \qquad L^b(z)a(w) \sim \frac{a(w)}{(z-w)^2} + \frac{\partial a(w)}{z-w} - \frac{2(a|b)k}{(z-w)^3}.$$

Hence (using (2.6.3)) we obtain:

$$(3.5.13) \qquad \left[L^b_m, a_n\right] = -na_{m+n} - (a|b)k(m^2 + m)\delta_{m,-n}.$$

In particular, $\left[L^b_{-1}, a_n\right] = -na_{n-1}$, hence $\left[L^b_{-1}, a(z)\right] = \partial a(z)$ and, as above, we deduce that $L^b(z)$ is a Virasoro field. Using (3.5.2), (3.5.8) and (3.5.11), we see that the central charge of $L^b(z)$ is equal to $\dim \mathfrak{h}_{\bar{0}} - \dim \mathfrak{h}_{\bar{1}} - 12(b|b)k$. Thus we have proved the following

PROPOSITION 3.5. *For each* $b \in \mathfrak{h}_{\bar{0}}$ *the field* $L^b(z)$ *is a Virasoro field with central charge*

$$(3.5.14) \qquad\qquad\qquad \operatorname{sdim} \mathfrak{h} - 12(b|b)k.$$

We apply now formula (3.5.10) to representation theory of the algebra $\hat{\mathfrak{h}}$. Since $\hat{\mathfrak{h}}$ is a direct sum of the abelian Lie superalgebra $\mathfrak{h}$ and the Heisenberg superalgebra

$$\hat{\mathfrak{h}}' = \bigoplus_{n \neq 0} \mathfrak{h} \otimes t^n + \mathbb{C}K,$$

it suffices to study representations of the latter. It is a $\mathbb{Z}$-graded Lie superalgebra with the triangular decomposition:

$$\hat{\mathfrak{h}}' = \hat{\mathfrak{h}}^< + \mathbb{C}K + \hat{\mathfrak{h}}^>, \quad \text{where } \hat{\mathfrak{h}}^{\lessgtr} = \bigoplus_{n > 0} \left(\mathfrak{h} \otimes t^{\mp n}\right).$$

The following lemma is immediate from the definitions.

LEMMA 3.5. *If* v *is a singular vector of a field representation of* $\hat{\mathfrak{h}}$ *(i.e.,* $\hat{\mathfrak{h}}^> v = 0$*), then* $Hv = 0$. $\qquad\qquad\square$

Let $\hat{\mathfrak{h}}^{\geqq} = \hat{\mathfrak{h}}^> + \mathbb{C}K$. Given $k \in \mathbb{C}$, denote by π^k the 1-dimensional representation of $\hat{\mathfrak{h}}^{\geqq}$ defined by:

$$\pi^k\left(\hat{\mathfrak{h}}^>\right) = 0, \qquad \pi^k(K) = k.$$

Then the Verma module $\widetilde{V}^k := \widetilde{V}(\pi^k)$ is explicitly described as follows:

$$\widetilde{V}^k = S(\hat{\mathfrak{h}}^<),$$

(i.e., $\widetilde{V}^k$ is identified with the symmetric superalgebra over the superspace $\hat{\mathfrak{h}}^<$), $K = kI$, $a_m = a \otimes t^m$ acts on $\widetilde{V}^k$ by multiplication if $m < 0$ and by a derivation of the symmetric superalgebra defined by

$$a_m(b \otimes t^{-n}) = km\delta_{m,n}(a|b), \qquad n > 0,$$

if $m > 0$.

THEOREM 3.5. (a) *The $\hat{\mathfrak{h}}'$-module $\widetilde{V}^k$ is irreducible iff $k \neq 0$. ($\widetilde{V}^0$ has a unique maximal submodule J^0 such that $\widetilde{V}^0/J^0$ is the trivial 1-dimensional module.)*

(b) *Any bounded field representation of $\hat{\mathfrak{h}}'$ such that $K = kI$ with $k \neq 0$ is equivalent to a direct sum of copies of the representation $\widetilde{V}^k$.*

PROOF. If $k \neq 0$ then we can construct the operator H. Due to (3.5.10), H is diagonalizable on $\widetilde{V}^k$ with non-negative eigenvalues and the only vectors with a zero eigenvalue are multiples of $1 \in \widetilde{V}^k$. Hence, by Lemma 3.5, $\widetilde{V}^k$ is irreducible if $k \neq 0$. The case $k = 0$ is obvious.

In order to prove (b), consider a bounded field representation of $\hat{\mathfrak{h}}'$ in a vector space V and denote by V^0 the subspace of V consisting of singular vectors. Since V is a bounded representation, it is clear that $V^0 \neq 0$. Since V is a field representation with $k \neq 0$, we can construct the operator H on V. It follows from (a) that $U(\hat{\mathfrak{h}}')v$ is an irreducible module isomorphic to $\widetilde{V}^k$ if v is a non-zero vector from V^0. Hence

$$V' := U(\hat{\mathfrak{h}}')V^0$$

is a direct sum of copies of the representation $\widetilde{V}^k$. Note that, due to (3.5.10), all eigenvalues of H on V' are non-negative.

Suppose now that $V \neq V'$. Then V/V' is again a bounded field $\hat{\mathfrak{h}}'$-module, hence there exists a non-zero singular vector $\bar{v} \in V/V'$, hence by Lemma 3.5, $H\bar{v} = 0$. Taking a preimage $v \in V$ of $\bar{v}$, which is an eigenvector of H, we obtain $Hv = 0$ and we see by the construction that $a_n v$ is a non-zero vector of V' for some $a \in \mathfrak{h}$ and some $n > 0$. Hence, by (3.5.10),

$$Ha_n v = -na_n v + a_n Hv = -na_n v.$$

Thus, $a_n v$ is an eigenvector of H in V' with a negative eigenvalue, a contradiction proving (b). $\square$

The $\hat{\mathfrak{h}}'$-module $B := \widetilde{V}^1$ is called the *oscillator representation* of the Heisenberg superalgebra $\hat{\mathfrak{h}}'$. It is characterized by the property of having a cyclic vector $|0\rangle = 1 \in B$ (i.e., $U\left(\hat{\mathfrak{h}}'\right)|0\rangle = B$) such that

$$(3.5.15) \qquad\qquad a_n|0\rangle = 0 \text{ for all } n > 0,\ a \in \mathfrak{h}.$$

EXAMPLE 3.5. The oscillator algebra $\mathfrak{s}$ (see (2.5.1)) is a special case when $\mathfrak{h} = \mathfrak{h}_{\bar{0}} = \mathbb{C}$, $(a|b) = ab$ and $\alpha_n = 1_n$. In this case the $\mathfrak{s}'$-module $\widetilde{V}^k$ can be identified with the algebra of polynomials $\mathbb{C}[x_1, x_2, \dots]$ so that $(m > 0)$:

$$\alpha_m = \frac{\partial}{\partial x_m}, \quad \alpha_{-m} = kmx_m, \quad K = k.$$

The (even) field $\alpha(z) = \sum_{n \in \mathbb{Z}} \alpha_n z^{-n-1}$ is local with the OPE

$$\alpha(z)\alpha(w) \sim \frac{k}{(z-w)^2}.$$

The $\mathfrak{s}'$-module $\widetilde{V}^k$ extends to an $\mathfrak{s}$-module $\widetilde{V}^{k;\mu}$ by letting $\alpha_0 = \mu \in \mathbb{C}$. Due to Theorem 3.5 any bounded field representation of $\mathfrak{s}$ such that $K = kI$ with $k \neq 0$ and α_0 is diagonalizable decomposes in a direct sum of representations $\widetilde{V}^{k;\mu}$, $\mu \in \mathbb{C}$. In particular, for each μ there exists a unique such irreducible representation.

It is easy to construct some "pathological" representations of $\mathfrak{s}$. If we take a 1-dimensional $\mathfrak{s}^{\geq}$-module π_1 such that $\alpha_n \mapsto 0$ for $n \gg 0$, then $\mathrm{Ind}_{\mathfrak{s}^{\geq}}^{\mathfrak{s}} \pi_1$ is a field representation which is not bounded. If we take a 2-dimensional representation π_2 of $\mathfrak{s}^{\geq}$ given by $\alpha_n \mapsto \left(\begin{smallmatrix} 0 & 0 \\ 1 & 0 \end{smallmatrix}\right)$ for all $n \geq 0$, $K \mapsto kI$, then $\mathrm{Ind}_{\mathfrak{s}^{\geq}}^{\mathfrak{s}} \pi_2$ is a bounded but not a field representation. It contains a submodule isomorphic to $\widetilde{V}^{k;0}$ the quotient by which is again isomorphic to $\widetilde{V}^{k;0}$, but the whole module is not $\widetilde{V}^{k;0} \oplus \widetilde{V}^{k;0}$.

3.6. Free (super)fermions

Now we consider the Clifford affinization of a finite-dimensional superspace A with non-degenerate anti-supersymmetric bilinear form $(.|.)$. Recall (see Section 2.5) that this is a Lie superalgebra

$$C_A = A[t, t^{-1}] + \mathbb{C}K$$

with commutation relations $(m, n \in \frac{1}{2} + \mathbb{Z}; \varphi, \psi \in A)$:

$$(3.6.1) \qquad [\varphi_m, \psi_n] = (\varphi|\psi)\delta_{m,-n} K, \quad [C_A, K] = 0,$$

where φ_m stands for $\varphi \otimes t^{m-\frac{1}{2}}$. Recall that the supercurrents

$$\varphi(z) = \sum_{n \in \frac{1}{2} + \mathbb{Z}} \varphi_n z^{-n-\frac{1}{2}}, \qquad \varphi \in A,$$

are mutually local with the OPE (2.5.8).

Consider a field representation of the Lie superalgebra C_A in a vector space V such that $K = kI_V$. We shall assume that $k \neq 0$. Then we obtain a set of mutually local field with the OPE

$$(3.6.2) \qquad \varphi(z)\psi(w) \sim \frac{(\varphi|\psi)k}{z - w},$$

called a system of *free fermions* (sometimes called superfermions if $A_{\bar{0}} \neq 0$). Note that these fields satisfy the conditions of Wick's theorem.

Choose dual bases $\{\varphi^i\}$ and $\{\psi^i\}$ of A (see Section 3.5), and consider the following even field of conformal weight 2:

$$(3.6.3) \qquad L(z) = \frac{1}{2k} \sum_i : \partial\varphi^i(z)\psi^i(z) := \sum_{n \in \mathbb{Z}} L_n z^{-n-2}.$$

Using Wick's theorem, we obtain

$$L(z)\varphi(w) \sim \frac{1}{2}\left(\frac{\varphi(z)}{(z - w)^2} + \frac{\partial\varphi(z)}{z - w}\right), \qquad \varphi \in A.$$

Hence, by Taylor's formula, we have:

$$(3.6.4) \qquad L(z)\varphi(w) \sim \frac{\frac{1}{2}\varphi(w)}{(z - w)^2} + \frac{\partial\varphi(w)}{z - w}.$$

Due to Table OPE (Section 2.6), this is equivalent to

$$(3.6.5) \qquad [L_m, \varphi_n] = -\left(\frac{1}{2}m + n\right)\varphi_{m+n}, \qquad m \in \mathbb{Z}, \quad n \in \frac{1}{2} + \mathbb{Z}.$$

The case $m = 0$ of (3.6.5) gives

$$(3.6.6) \qquad [L_0, \varphi_n] = -n\varphi_n, \qquad n \in \frac{1}{2} + \mathbb{Z}, \quad \varphi \in A,$$

i.e., $\varphi(z)$ has conformal weight $\frac{1}{2}$ with respect to the Hamiltonian $\mathrm{ad}L_0$. The case $m = -1$ of (3.6.5) gives:

$$[L_{-1}, \varphi(z)] = \partial\varphi(z), \qquad \varphi \in A.$$

In the same way as for free bosons, it follows that $L(z)$ is a Virasoro field. Computing the $s = 2$ term of $L(z)L(w)$ in Wick's formula, we obtain

$$(3.6.7) \qquad \text{central charge of } L(z) = -\frac{1}{2}\,\text{sdim}\,A.$$

In the same way as in the bosonic case, we apply (3.6.6) to representation theory of the Lie superalgebra C_A. It is a $\mathbb{Z}$-graded (by adL_0) Lie superalgebra with the triangular decomposition:

$$C_A = C_A^< + \mathbb{C}K + C_A^>,$$

where $C_A^> = A \otimes \mathbb{C}[t]$, $C_A^< = A \otimes t^{-1}\mathbb{C}[t^{-1}]$. Let $C_A^\geq = C_A^> + \mathbb{C}K$. Given $k \in \mathbb{C}$ denote by π^k the 1-dimensional representation of $C_A^\geq$ defined by $\pi^k(C_A^>) = 0$, $\pi^k(K) = k$. Then the Verma module $\widetilde{V}^k := \widetilde{V}(\pi^k)$ is identified with $S(C_A^<)$, $K = kI$, φ_m acts by multiplication if $m < 0$ and by a derivation of the superalgebra $S(C_A^<)$ defined by

$$\varphi_m\left(\psi \otimes t^{-n}\right) = k\delta_{m,-n+1}(\varphi|\psi), \qquad n > 0,$$

if $m \geq 0$.

The following result is proved in exactly the same way as Theorem 3.5, by making use of (3.6.6).

THEOREM 3.6. (a) *The C_A-module $\widetilde{V}^k$ is irreducible iff $k \neq 0$.*

(b) *Any bounded field representation of C_A such that $K = kI$ with $k \neq 0$ is equivalent to a direct sum of copies of the representation $\widetilde{V}^k$.*

The C_A-module $F := \widetilde{V}^1$ is called the *spin representation* of the Clifford Lie superalgebra C_A. It is characterized by the property of having a cyclic vector $|0\rangle \in F$ such that

$$(3.6.8) \qquad \varphi_n|0\rangle = 0 \quad \text{for all } n > 0, \quad \varphi \in A.$$

In conclusion of this section we describe a very useful construction, called *bosonization.* Suppose that the superspace A is a direct sum of two isotropic subspaces A^+ and A^-, and let $k = 1$. Choose bases $\{\varphi^i\}$ of A^+ and $\{\psi^i\}$ of A^- such that $(\psi^i|\varphi^j) = \delta_{ij}$. Note that for any $\varphi \in A$ we have

$$(3.6.9) \qquad \varphi = \varphi^+ + \varphi^-, \quad \text{where } \varphi^+ = \sum_i (\psi^i|\varphi)\,\varphi^i, \quad \varphi^- = \sum_i (\varphi|\varphi^i)\,\psi^i.$$

Construct a new field of conformal weight 1:

$$\alpha(z) = \sum_i \; : \varphi^i(z)\psi^i(z) : \; .$$

Using Wick's and Taylor formulas, formulas (3.6.2) for $k = 1$ and (3.6.9) we obtain the following OPE:

$$(3.6.10) \qquad \alpha(z)\varphi(w) \sim \frac{\varphi^+(w) - \varphi^-(w)}{z - w}.$$

Furthermore, Wick's formula gives:

$$(3.6.11) \qquad \alpha(z)\alpha(w) \;=\; -\frac{\mathrm{sdim}\, A^+}{(z-w)^2} + \frac{\displaystyle\sum_i (: \varphi^i(z)\psi^i(w) : - : \varphi^i(w)\psi^i(z) :)}{z - w}$$
$$+ \sum_{i,j} \; : \varphi^i(z)\psi^i(z)\varphi^j(w)\psi^j(w) : \; .$$

By Taylor's formula, the second term on the right-hand side of (3.6.11) equals

$$\sum_i (: \partial\varphi^i(w)\psi^i(w) : - : \varphi^i(w)\partial\psi^i(w) :) + (z - w)(\cdots),$$

and the third term equals

$$\sum_{i,j} \; : \varphi^i(w)\psi^i(w)\varphi^j(w)\psi^j(w) : + (z - w)(\cdots).$$

We conclude that

$$(3.6.12) \qquad \alpha(z)\alpha(w) \sim \frac{-\,\mathrm{sdim}\, A^+}{(z-w)^2},$$

i.e., that $\alpha(z)$ is a free boson with affine central charge $-\,\mathrm{sdim}\, A^+$, and that

$$: \alpha(w)\alpha(w) : \;=\; \sum_i \left(: \partial\varphi^i(w)\psi^i(w) : - : \varphi^i(w)\partial\psi^i(w) :\right)$$
$$(3.6.13) \qquad\qquad + \sum_{i,j} \; : \varphi^i(w)\psi^i(w)\varphi^j(w)\psi^j(w) : \; .$$

Finally, note that we may construct a family of Virasoro fields

$$(3.6.14) \qquad L^\lambda(z) = (1 - \lambda)L^+(z) + \lambda L^-(z), \quad \lambda \in \mathbb{C},$$

where

$$L^+(z) = \sum_i \; : \partial\varphi^i(z)\psi^i(z) : , \quad L^-(z) = \sum_i \; : \partial\psi^i(z)\varphi^i(z) : ,$$

so that using Wick and Taylor formulas we obtain:

$$(3.6.15) \qquad L^\lambda(z)\varphi(w) \sim \frac{\partial\varphi(w)}{z-w} + \frac{(1-\lambda)\varphi^-(w) + \lambda\varphi^+(w)}{(z-w)^2}, \qquad \varphi \in A.$$

It follows as above that $L^\lambda(z)$ are Virasoro fields for each λ. The central charge, calculated as before, is equal to

$$(3.6.16) \qquad c_\lambda = (12\lambda^2 - 12\lambda + 2)\,\mathrm{sdim}\,A^+.$$

CHAPTER 4

Structure theory of vertex algebras

4.1. Consequences of translation covariance and vacuum axioms

First, recall the axioms of a vertex algebra given in Section 1.3. It is often convenient to state them in a slightly different form (closer in spirit to the Wightman axioms).

A vertex algebra is a superspace V endowed with a vector $|0\rangle$ (vacuum vector), an endomorphism T (infinitesimal translation operator) and a parity preserving linear map of V to the space of fields (the state-field correspondence)

$$a \mapsto Y(a, z) = \sum_{n \in \mathbb{Z}} a_{(n)} z^{-n-1}, \qquad a_{(n)} \in \mathrm{End} V,$$

such that the following axioms hold $(a, b \in V)$:

(translation covariance): $[T, Y(a, z)] = \partial Y(a, z)$,

(vacuum): $T|0\rangle = 0$, $Y(|0\rangle, z) = I_V$, $Y(a, z)|0\rangle|_{z=0} = a$,

(locality): $(z - w)^N [Y(a, z), Y(b, w)] = 0$ for $N \gg 0$.

Applying both sides of the translation invariance axiom to $|0\rangle$ we obtain (1.3.3) from the 1st and 3rd parts of the vacuum axiom after letting $z = 0$. Hence these axioms imply those in Section 1.3. Conversely, $T|0\rangle = 0$ follows from (1.3.3) and the 2nd part of the vacuum axiom.

The fields $Y(a, z)$ are often called *vertex operators*, hence the name vertex algebra.

The following theorem provides a general, albeit abstract, construction of vertex algebras (cf. [**Li**]).

THEOREM 4.1. *Any local linear field algebra $F \subset glf(U)$ is a vertex algebra with the vacuum vector $|0\rangle = I_U$, the infintesimal translation operator $T = \partial_z$ and the vertex operators*

$$Y(a(z), x)b(z) = \sum_{n \in \mathbb{Z}} (a(z)_{(n)} b(z)) x^{-n-1}.$$

PROOF. First, the vertex operators $Y(a(z), x)$ are $\mathrm{End}F$-valued fields since F consists of $(\mathrm{End}U$-valued) mutually local fields. Moreover, these vertex operators are pairwise local due to Proposition 3.2. This checks the locality axiom. The vacuum axioms mean the following:

$$\partial_z I_U = 0, \, I_{U_{(n)}} a(z) = \delta_{n,-1} a(z) \text{ for } n \in \mathbb{Z}, \, a(z)_{(n)} I_U = \delta_{n,-1} a(z) \text{ for } n \geq -1.$$

These formulas are obvious. The translation covariance axiom means:

$$[\partial_z, \, a(z)_{(n)}] b(z) = -n a(z)_{(n-1)} b(z) \text{ for } n \in \mathbb{Z}.$$

But this is formula (3.3.5b).

$\square$

The following easy uniqueness (and existence) theorem for a formal differential equation is very useful in establishing identities.

LEMMA 4.1. *Let U be a vector space and let $R(z) \in (\mathrm{End}U)[[z]]$. Then the differential equation*

$$(4.1.1) \qquad\qquad \frac{d}{dz} f(z) = R(z) f(z)$$

has a unique solution of the form

$$f(z) = \sum_{n \in \mathbb{Z}_+} f_n z^n, \qquad f_n \in U,$$

with the given initial data f_0.

PROOF. Equation (4.1.1) means:

$$j f_j = \sum_{i=0}^{j-1} R_i f_{j-i-1} \quad \text{for} \quad j \geq 1, \quad \text{where } R(z) = \sum_{j \in \mathbb{Z}_+} R_j z^j.$$

These equations allow one to compute the f_j, $j > 0$, recursivity for each given f_0.

$\square$

PROPOSITION 4.1. *(a) For any element a of a vertex algebra V one has*

$$(4.1.2) \qquad\qquad Y(a, z)|0\rangle \;=\; e^{zT}(a),$$

$$(4.1.3) \qquad e^{zT} Y(a, w) e^{-zT} \;=\; Y(a, z + w) \text{ in the domain } |z| < |w|,$$

$$(4.1.4) \qquad e^{zT} Y(a, w)_{\pm} e^{-zT} \;=\; Y(a, z + w)_{\pm} \text{ in the domain } |z| < |w|.$$

(b) For any two elements $a, b \in V$ and any $n \in \mathbb{Z}$ one has:

$$(4.1.5) \qquad Y(a_{(n)}b, z)|0\rangle = (Y(a, z)_{(n)}Y(b, z))|0\rangle .$$

PROOF. We actually proved already (4.1.2) in Section 1.3. It is placed here again because the proof of all four formulas is the same.

Note that (4.1.2) and (4.1.5) are equalities in $V[[z]]$ and (4.1.3 and 4) are equalities in $\mathrm{End}V\left[[w, w^{-1}]\right][[z]]$ (recall that "in the domain $|z| < |w|$" means that $(z + w)^j$ is replaced by its power series expansion $i_{w,z}(z + w)^j \in \mathbb{C}\left[[w, w^{-1}]\right][[z]]$).

In order to prove (4.1.2) and (4.1.5), we apply Lemma 4.1 to $U = V$, $R = T$. Since both sides of (4.1.2) obviously satisfy the differential equation (4.1.1) with the initial condition $f_0 = a$, (4.1.2) follows. Both sides of (4.1.5) satisfy the differential equation (4.1.1) due to the vacuum and translation covariance axioms and the fact that both ∂_z and $\mathrm{ad}T$ are derivations of all n-th products (see Proposition 3.3(a) and (3.1.12)). The coincidence of the intial conditions follows from the vacuum axiom and Lemma 3.1.

In order to prove the remaining two formulas, we apply Lemma 4.1 to $U = (\mathrm{End}V)\left[[w, w^{-1}]\right]$, $R = \mathrm{ad}T$. Since both sides of (4.1.3) (resp. (4.1.4)) satisfy (4.1.1) with the initial condition $f_0 = Y(a, w)$ (resp. $f_0 = Y(a, w)_{\pm}$), (4.1.3 and 4) follow. In the proof of (4.1.4) we have used that the translation covariance equation splits into two equations:

$$(4.1.6) \qquad [T, Y(a, z)_{\pm}] = \partial Y(a, z)_{\pm}.$$

$\square$

4.2. Skewsymmetry

PROPOSITION 4.2. *For any elements a and b of a vertex algebra V one has the following skewsymmetry relation:*

$$(4.2.1) \qquad Y(a, z)b = (-1)^{p(a)p(b)}e^{zT}Y(b, -z)a.$$

PROOF. We have by the locality axiom for $N \gg 0$:

$$(z - w)^N Y(a, z)Y(b, w)|0\rangle = (-1)^{p(a)p(b)}(z - w)^N Y(b, w)Y(a, z)|0\rangle.$$

This can be rewritten using (4.1.2):

$$(z - w)^N Y(a, z)e^{wT}b = (-1)^{p(a)p(b)}(z - w)^N Y(b, w)e^{zT}a.$$

Applying (4.1.3) to the right-hand side we get

$$(4.2.2) \qquad (z-w)^N Y(a,z) e^{wT} b = (-1)^{p(a)p(b)} (z-w)^N e^{zT} Y(b, i_{w,z}(w-z))a.$$

Since $b_{(n)}(a) = 0$ (resp. $a_{(n)}(b) = 0$) for $n \gg 0$, the equality (4.2.2) involves only positive powers of $z-w$ if N is sufficiently large (resp. only finitely many negative powers of z). Hence (4.2.2) is an equality in $(\mathrm{End}\,V)((z))\,[[z-w]]$ if N is sufficiently large. Then we can let $w=0$ in both sides of (4.2.2) and divide by z^N, obtaining (4.2.1). $\qquad\qquad\square$

Comparing coefficients of (4.2.1) we obtain the original Borcherds formula [**B1**] for skewsymmetry ($n \in \mathbb{Z}$):

$$(4.2.3) \qquad\qquad a_{(n)}b = -p(a,b) \sum_{j=0}^{\infty} (-1)^{j+n} T^{(j)} \left(b_{(n+j)}a \right).$$

Here and further we write $a_{(n)}b$ in place of $a_{(n)}(b)$ (the endomorphism $a_{(n)}$ applied to a vector b). We do this not only for typographical reasons, but, more importantly, in order to emphasize that for each $n \in \mathbb{Z}$ we have on V a $\mathbb{C}$-bilinear product $a_{(n)}b$. As we shall see, the products $a_{(n)}b$ are essentially the same as products $a(z)_{(n)}b(z)$ discussed in Sections 2.3 and 3.1. Formula (4.2.3) is the counterpart of Proposition 3.3(b).

REMARK 4.2. Theorem 4.1 and formula (4.2.3) give another proof of Proposition 3.3(b). Indeed, just consider the linear field algebra F generated by the fields $a(z)$ and $b(z)$.

4.3. Subalgebras, ideals, and tensor products

A *subalgebra* of a vertex algebra V is a subspace U of V containing $|0\rangle$ such that

$$a_{(n)}U \subset U \quad \text{for all } a \in U.$$

It is clear that U is a vertex algebra too, its fields being $Y(a,z) = \sum_n a_{(n)}|_U z^{-n-1}$. This follows immediately from the axioms of a vertex algebra in Section 1.3.

A *homomorphism* of a vertex algebra V to a vertex algebra V' is a linear parity preserving map $\varphi : V \to V'$ such that

$$\varphi(a_{(n)}b) = \varphi(a)_{(n)}\varphi(b) \quad \text{for all} \quad a,b \in V, \quad n \in \mathbb{Z}.$$

A *derivation* D of parity $\gamma \in \mathbb{Z}/2\mathbb{Z}$ of a vertex algebra V is an endomorphism of the space V such that $DV_\alpha \subset V_{\alpha+\gamma}$ and

$$D(a_{(n)}b) = (Da)_{(n)}b + (-1)^{\alpha\gamma}a_{(n)}(Db) \quad \text{for all} \quad a \in V_\alpha, \quad b \in V.$$

Note that if D is an even derivation and e^D is a convergent series, then e^D is an automorphism of the vertex algebra V.

An *ideal* of a vertex algebra V is a T-invariant subspace J not containing $|0\rangle$ such that

$$a_{(n)}J \subset J \text{ for all } a \in V.$$

Note that we have

$$(4.3.1) \qquad\qquad a_{(n)}V \subset J \text{ for all } a \in J.$$

Indeed, it follows from the skewsymmetry (4.2.1) that $Y(a,z)v = \pm e^{zT}Y(v,-z)a \in J[[z,z^{-1}]]$ for $a \in J$, $v \in V$. Hence the quotient space V/J has a canonical structure of a vertex algebra, and we have a canonical homomorphism $V \to V/J$ of vertex algebras.

The *tensor product* of two vertex algebras U and V is defined as follows. The space of states is $U \otimes V$, the vacuum vector is $|0\rangle \otimes |0\rangle$, the infinitesimal translation operator is $T \otimes 1 + 1 \otimes T$. Finally, the fields are

$$Y(u \otimes v, z) = Y(u,z) \otimes Y(v,z) \equiv \sum_{m,n\in\mathbb{Z}} u_{(m)} \otimes v_{(n)} z^{-m-n-2}.$$

In other words

$$(4.3.2) \qquad\qquad (u \otimes v)_{(k)} = \sum_{m\in\mathbb{Z}} u_{(m)} \otimes v_{(-m+k-1)}.$$

We use the usual definition of a tensor product of two operators A and B:

$$(A \otimes B)(a \otimes b) = (-1)^{p(B)p(a)} A(a) \otimes B(b).$$

It is clear that the sum (4.3.2) applied to any vector $a \otimes b$ is finite (since both $Y(u,z)$ and $Y(v,z)$ are fields). We have that $(u\otimes v)_{(k)}(a\otimes b) = 0$ for $k \gg 0$ because $u_{(m)}a = 0$ for $m \geq M$ and $v_{(n)}b = 0$ for $n \geq N$ imply $u_{(m)} \otimes v_{(-m+k-1)}(a \otimes b) = 0$ for $k > M + N$.

It is straightforward to check that $U \otimes V$ is a vertex algebra.

Given a vertex algebra V it is natural to define its *affinization* $\hat{V}$ as follows [**B1**]. Let $\mathbb{C}\left[t, t^{-1}\right]$ be the algebra of Laurent polynomials (with trivial $\mathbb{Z}/2\mathbb{Z}$-gradation) and let T denote its derivation ∂_t. Then $\mathbb{C}\left[t, t^{-1}\right]$ is endowed with the structure of a holomorphic vertex algebra (see Section 1.4), and we let

$$\hat{V} = \mathbb{C}\left[t, t^{-1}\right] \otimes V.$$

Of course, this affinization is closely related to that considered in Section 2.7.

4.4. Uniqueness theorem

The following uniqueness theorem is extremely useful in identifying a field with one of the fields of a vertex algebra.

THEOREM 4.4. [**G**] *Let V be a vertex algebra and let $B(z)$ be a field (with values in* $\mathrm{End}V$*) which is mutually local with all the fields* $Y(a, z)$*,* $a \in V$*. Suppose that for some* $b \in V$*:*

$$(4.4.1) \qquad\qquad B(z)|0\rangle = e^{zT}b.$$

Then $B(z) = Y(b, z)$*.*

PROOF. By the assumption of locality we have:

$$(z - w)^N B(z)Y(a, w)|0\rangle = (-1)^{p(B)p(a)}(z - w)^N Y(a, w)B(z)|0\rangle.$$

Applying to the left- (resp. right) hand side formula (4.1.2) (resp. (4.4.1)) we obtain:

$$(4.4.2) \qquad (z - w)^N B(z)e^{wT}a = (-1)^{p(B)p(a)}(z - w)^N Y(a, w)e^{zT}b.$$

Applying (4.1.2) to the right-hand side of (4.4.2) we get

$$(-1)^{p(B)p(a)}(z - w)^N Y(a, w)Y(b, z)|0\rangle$$

which by locality (for sufficiently large N) is equal to $(z - w)^N Y(b, z)Y(a, w)|0\rangle$. (It follows from (4.4.1) that $p(B) = p(b)$ since $p(T) = 0$.) Applying to this (4.1.2) again and equating it with the left-hand side of (4.4.2), we obtain

$$(z - w)^N B(z)e^{wT}a = (z - w)^N Y(b, z)e^{wT}a.$$

Letting $w = 0$ and dividing by z^N, we get $B(z)a = Y(b, z)a$ for any $a \in V$. $\square$

REMARK 4.4a. Condition (4.4.1) follows from

$$(4.4.3) \qquad B(z)|0\rangle|_{z=0} = b, \quad \partial B(z)|0\rangle = TB(z)|0\rangle.$$

Indeed, equation (4.4.3) means that $B(z)|0\rangle$ is a solution of the differential equation $\frac{d}{dz}a(z) = Ta(z)$, $a(z) \in V[[z]]$, with the initial condition $a_0 = b$. Due to Lemma 4.1 we conclude that (4.4.1) holds.

Note that just the first of the condition (4.4.3) is not enough as the example $B(z) = (1 + z)Y(b, z)$ shows.

The first corollary of the Uniqueness theorem is the following important proposition.

PROPOSITION 4.4. *For any two elements a and b of a vertex algebra V and any $n \in \mathbb{Z}$ one has:*

$$(4.4.4) \qquad Y(a_{(n)}b, z) = Y(a, z)_{(n)}Y(b, z).$$

PROOF. Let $B(z) = Y(a, z)_{(n)}Y(b, z)$. By (4.1.5) and (4.1.2) we have:

$$B(z)|0\rangle = Y(a_{(n)}b, z)|0\rangle = e^{zT}(a_{(n)}b).$$

Since, by Dong's lemma, $B(z)$ is local with respect to all vertex operators $Y(c, z)$, (4.4.4) follows from Theorem 4.4. $\qquad\square$

COROLLARY 4.4. (a) *In a vertex algebra V for any collection of vectors $a^1, \ldots, a^n \in V$ and any collection of positive integers $j_1, \ldots, j_k$ one has*

$$(4.4.5) \qquad : \partial^{(j_1-1)}Y(a^1, z) \cdots \partial^{(j_n-1)}Y(a^n, z) := Y\left(a^1_{(-j_1)} \cdots a^n_{(-j_n)}|0\rangle, z\right).$$

(b) *For any $a, b \in V$ and any $n \in \mathbb{Z}_+$ one has:*

$$(4.4.6) \qquad : \partial^{(n)}Y(a, z)Y(b, z) := Y\left(a_{(-n-1)}b, z\right).$$

(c) *For any $a \in V$ one has*

$$(4.4.7) \qquad Y(Ta, z) = \partial Y(a, z).$$

PROOF. (a) and (b) follow from Proposition 4.4 due to (3.1.6). Since $Ta = a_{(-2)}|0\rangle$, (c) is a special case of (a) when $n = 1$ and $j_1 = 2$. $\qquad\square$

REMARK 4.4b. Let $\operatorname{Vac} V = \{a \in V | Ta = 0\}$. This subspace contains $\mathbb{C}|0\rangle$ but may be larger (see Remark 5.7c). (One often imposes an additional axiom of QFT requiring uniqueness of the vacuum, but we do not require this). It follows from (4.4.7) that

$$\operatorname{Vac} V = \{a \in V \mid Y(a,z) = a_{(-1)}\}$$

and from (4.4.6) that $\operatorname{Vac} V$ is a subalgebra of V. This is called the *vacuum subalgebra* of the vertex algebra V. It follows from locality that

(4.4.8) $$\left[a_{(-1)}, Y(b,z)\right] = 0 \text{ for } a \in \operatorname{Vac} V, \quad b \in V.$$

Hence

(4.4.9) $$b_{(n)} \operatorname{Vac} V = 0 \text{ for } b \in V, n \in \mathbb{Z}_+.$$

4.5. Existence theorem

The following theorem allows one to construct vertex algebras (cf. [**FKRW**]).

THEOREM 4.5. *Let V be a vector superspace, let $|0\rangle$ be an even vector of V and T an even endomorphism of V. Let $\{a^\alpha(z)\}_{\alpha \in A}$ (A an index set) be a collection of fields such that*

(i) $[T, a^\alpha(z)] = \partial a^\alpha(z)$ $(\alpha \in A)$,

(ii) $T|0\rangle = 0$, $a^\alpha(z)|0\rangle|_{z=0} = a^\alpha$ $(\alpha \in A)$, $\quad |0\rangle \notin \sum_\alpha \mathbb{C}a^\alpha$,

(iii) *the linear map:* $\sum_\alpha \mathbb{C}a^\alpha(z) \to \sum_\alpha \mathbb{C}a^\alpha$, *defined by* $a^\alpha(z) \mapsto a^\alpha$, *is injective,*

(iv) $a^\alpha(z)$ *and* $a^\beta(z)$ *are mutually local* $(\alpha, \beta \in A)$,

(v) *the vectors* $a^{\alpha_1}_{(j_1)} \cdots a^{\alpha_n}_{(j_n)}|0\rangle$ *with* $j_s \in \mathbb{Z}, \alpha_s \in A$ *span* V.

Then the formula

(4.5.1) $$Y\left(a^{\alpha_1}_{(j_1)} a^{\alpha_2}_{(j_2)} \cdots a^{\alpha_n}_{(j_n)}|0\rangle, z\right) = a^{\alpha_1}(z)_{(j_1)}(a^{\alpha_2}(z)_{(j_2)}(\cdots(a^{\alpha_n}(z)_{(j_n)}I_V)))$$

defines a unique structure of a vertex algebra on V such that $|0\rangle$ is the vacuum vector, T is the infinitesimal translation operator and

(4.5.2) $$Y(a^\alpha, z) = a^\alpha(z), \quad \alpha \in A.$$

PROOF. Choose a basis among the vectors of the form (v) which contains $|0\rangle$ and a basis of $\sum \mathbb{C}a^\alpha$, and define $Y(a,z)$ by formula (4.5.1). By (iv), Remark 2.3a and Dong's lemma, the locality axiom holds. It follows from Lemma 3.1 and (ii),

(iii) that the vacuum axioms hold (the first two of them hold for trivial reasons). Finally, the operators $\mathrm{ad}T$ and ∂ are both derivations of the n-th products (see Proposition 3.3.(a) and (3.1.2)), which, due to (i), coincide on the $a^\alpha(z)$. The translation covariance axiom follows.

Thus we get a structure of a vertex algebra on V which satifies (4.5.2). Formula (4.5.1) holds for all vectors of the form (v) due to (4.4.4). Uniqueness is clear as well, since (4.5.1) must hold **due** to (4.4.4).

$\square$

DEFINITION 4.5. A collection of fields of a vertex algebra V satisfying condition (v) of Theorem 4.5 is called a *generating set of fields* of V. If condition (v) holds with the additional assumption that all $j_s < 0$, this collection is called a *strongly generating set of fields*.

4.6. Borcherds OPE formula

Let V be a vertex algebra. We have:

$$Y(a,z)Y(b,w)|0\rangle = Y(a,z)e^{wT}b = e^{wT}Y(a,z-w)b$$

(the last equality holds in the domain $|z| > |w|$ due to (4.1.4)). Letting $c = Y(a,z-w)b$ we have

$$Y(a,z)Y(b,w)|0\rangle = e^{wT}c.$$

If the uniqueness theorem were applicable we would derive the "associativity" of V:

$$(4.6.1a) \qquad Y(a,z)Y(b,w) = Y(Y(a,z-w)b,w)$$

$$(4.6.1b) \qquad\qquad\qquad = \sum_{n\in\mathbb{Z}} \frac{Y(a_{(n)}b,w)}{(z-w)^{n+1}},$$

the latter equality being the "symbolic" OPE. However, the uniqueness theorem is not quite applicable (and no wonder, since the "symbolic" OPE makes no sense as an equality of formal distributions). In the "graded" case this "proof" can be made rigorous by making use of the analytic continuation (cf. Remark 4.9a). Still, in view of the discussion in Section 3.1, we may expect that the following holds.

THEOREM 4.6. *In the domain $|z| > |w|$ one has for any $a, b \in V$:*

$$(4.6.2a) \qquad Y(a,z)Y(b,w) = \sum_{n=0}^{\infty} \frac{Y\left(a_{(n)}b,w\right)}{(z-w)^{n+1}} + :Y(a,z)Y(b,w):\,.$$

Equivalently:

$$(4.6.2b) \qquad [Y(a,z),Y(b,w)] = \sum_{n=0}^{\infty} Y(a_{(n)}b,w)\partial_w^{(n)}\delta(z-w).$$

PROOF. We have by the OPE (2.3.9a):

$$[Y(a,z),Y(b,w)] = \sum_{n\in\mathbb{Z}_+} (Y(a,w)_{(n)}Y(b,w))\partial_w^{(n)}\delta(z-w).$$

Hence the theorem follows from Proposition 4.4. $\qquad\qquad\qquad\square$

Formula (4.6.2b) is equivalent to each of the following very useful Borcherds commutator formulas $(m,n\in\mathbb{Z})$:

$$(4.6.3) \qquad [a_{(m)},b_{(n)}] \;=\; \sum_{j\geq 0}\binom{m}{j}\left(a_{(j)}b\right)_{(m+n-j)}$$

$$(4.6.4) \qquad [a_{(m)},Y(b,z)] \;=\; \sum_{j\geq 0}\binom{m}{j}Y\left(a_{(j)}b,z\right)z^{m-j}.$$

REMARK 4.6a. With respect to $\partial = T$ and all products $a_{(n)}b$ for $n\in\mathbb{Z}_+$, a vertex algebra V is a conformal superalgebra. Furthermore, the linear map $\mathrm{Lie}\,V\to \mathrm{End}V$ defined by $a_n \mapsto a_{(n)}$, $a\in V$, $n\in\mathbb{Z}$, is a Lie superalgebra homomorphism. Its kernel is an irregular ideal of $\mathrm{Lie}\,V$. (Here $\mathrm{Lie}\,V$ is the maximal formal distribution Lie superalgebra associated to V viewed as a conformal superalgebra.) This is immediate by formulas (1.3.4), (4.2.3) and (4.6.3).

An important special case of (4.6.4) is

$$(4.6.5) \qquad\qquad [a_{(0)},Y(b,z)] = Y(a_{(0)}b,z).$$

COROLLARY 4.6. (a) $a_{(0)}b = 0$ *iff* $[a_{(0)},Y(b,z)] = 0$.

(b) $a_{(j)}b = 0$ *for all* $j\in\mathbb{Z}_+$ *iff* $[Y(a,z),Y(b,w)] = 0$.

(c) *The operator* $a_{(0)}$ *is a derivation of the vertex algebra* V *for any* $a\in V$. *These derivations form a subalgebra of the Lie superalgebra of all derivations of* V.

(d) *The centralizer of* $Y(a,z)$ *in* V *(i.e., the subspace* $\{b\in V \mid [Y(a,z),Y(b,w)] = 0\}$*) is a vertex subalgebra of* V.

(e) *A subspace* U *of* V *is a vertex subalgebra iff the collection of fields* $\{Y(a,z)\mid a\in U\}$ *is a linear field algebra.*

(f) *The fixed point set of an automorphism of* V *is a vertex subalgebra of* V.

(g) *If a vertex algebra V is generated by a collection of fields $Y(a^i, z)$ and $b, b' \in V$ are such that $b_{(0)} a^i = b'_{(0)} a^i$ for all a^i, then $b_{(0)} = b'_{(0)}$.*

(h) *If a vertex algebra V is generated by a collection of fields which is closed under OPE (i.e., all OPE coefficients are linear combinations of these fields or their derivatives), then V is strongly generated by this collection of fields.*

PROOF. (a) follows from (4.6.5) and (b) is immediate from (4.6.4). The first part of (c) follows from (4.6.3) for $m = 0$ applied to $c \in V$. The second part of (c) follows from (4.6.3) for $m = n = 0$. (d) follows from (b). (e) is clear by (4.4.4). (f) is obvious. (g) follows from (c). Finally, (h) follows from formula (4.6.3) which shows that the bracket $[a_{(m)}, b_{(n)}]$ with $m \geq 0$ and $n < 0$ is a linear combination of some $c_{(k)}$ with $k < m$, hence applying $a_{(m)}$ to an element of the form $a^{\alpha_1}_{(j_1)} \cdots a^{\alpha_n}_{(j_n)} |0\rangle$ with the $j_s < 0$, we get by induction a linear combination of elements of this form. $\square$

REMARK 4.6b. Corollary 4.6 provides several ways of constructing subalgebras of a vertex algebra V, which are quite popular in both mathematics and physics literature:

(I) Given a subspace U of V, its centralizer

$$C_V(U) = \{b \in V \,|\, [Y(a, z), Y(b, w)] = 0 \quad \text{for all} \quad a \in U\}$$

is a subalgebra of V (by Corollary 4.6d) called by physicists a coset model.

(II) Given a collection of elements $\{a^i\}$ of V, the intersection of the null spaces of the operators $a^i_{(0)}$ is a subalgebra of V (due to Corollary 4.6c) called by physicists a W-algebra.

(III) Given a collection of elements $\{a^i\}$ of V, the linear span of all the vectors

$$a^{i_1}_{(n_1)} \cdots a^{i_s}_{(n_s)} |0\rangle$$

is a subalgebra of V generated by the fields $Y(a^i, z)$.

(IV) Given a group of automorphisms G of a vertex algebra V, the fixed point set V^G is a subalgebra of V (by Corollary 4.6f), called by physicists an orbifold model when G is finite.

4.7. Vertex algebras associated to formal distribution Lie superalgebras

Let $\mathfrak{g}$ be a Lie superalgebra spanned by mutually local formal distributions $a^\alpha(z)$ $(\alpha \in A)$, and suppose that there exists an endomorphism T of the space $\mathfrak{g}$

over $\mathbb{C}$ such that

$$(4.7.1) \qquad Ta^\alpha(z) = \partial a^\alpha(z).$$

Then $\mathfrak{g}$ is called a *regular* formal distribution Lie superalgebra. It is clear that T is an even derivation of the Lie superalgebra $\mathfrak{g}$ given by the formula

$$(4.7.2) \qquad Ta^\alpha_{(n)} = -na^\alpha_{(n-1)} \,.$$

Note that $\mathrm{Lie}\,R$, where R is a conformal superalgebra (see Section 2.7), is regular, in particular current algebras and the Virasoro algebra are regular. Let

$$\mathfrak{g}_{--} = \left\{ a \in \mathfrak{g} \mid T^k a = 0 \text{ for } k \gg 0 \right\}.$$

This is a T-invariant subalgebra of $\mathfrak{g}$ which, due to (4.7.2), contains the annihilation subalgebra $\mathfrak{g}_-$ of $\mathfrak{g}$ (see (3.4.1)):

$$(4.7.3) \qquad \mathfrak{g}_{--} \supset \mathfrak{g}_- \,.$$

Let $\lambda : \mathfrak{g}_{--} \to \mathbb{C}$ be a 1-dimensional $\mathfrak{g}_{--}$-module such that

$$(4.7.4) \qquad \lambda(T\mathfrak{g}_{--}) = 0.$$

Consider the induced $\mathfrak{g}$-module (cf. Section 3.4)

$$(4.7.5) \qquad \widetilde{V}^\lambda(\mathfrak{g}) := \mathrm{Ind}^{\mathfrak{g}}_{\mathfrak{g}_{--}} \lambda = U(\mathfrak{g})/U(\mathfrak{g}) \langle a - \lambda(a) \mid a \in \mathfrak{g}_{--} \rangle,$$

and let $|0\rangle \in \widetilde{V}^\lambda(\mathfrak{g})$ be the image of $1 \in U(\mathfrak{g})$.

Note that the formal distributions $a^\alpha(z)$ are represented in $\widetilde{V}^\lambda(\mathfrak{g})$ by fields (which we shall denote by the same symbol). This follows from (4.7.3) and the discussion in Section 3.4.

The derivation T of $\mathfrak{g}$ extends to a derivation of $U(\mathfrak{g})$, which can be pushed down to an endomorphism of the space $\widetilde{V}^\lambda(\mathfrak{g})$ due to condition (4.7.4). This endomorphism is again denoted by T.

The following theorem is now an immediate corollary of the Existence Theorem 4.5.

THEOREM 4.7. *Let $\mathfrak{g}$ be a regular formal distribution Lie superalgebra. Then the $\mathfrak{g}$-module $\widetilde{V}^\lambda(\mathfrak{g})$ has a unique vertex algebra structure with $|0\rangle$ the vacuum vector and generated by the fields $a^\alpha(z)$ ($\alpha \in A$).*

REMARK 4.7. A formal distribution $a^\alpha(z)$ is represented in $\widetilde{V}^\lambda(\mathfrak{g})$ by a zero field iff $a^\alpha_{(-1)} \in \mathfrak{g}_{--}$. It follows from (4.7.2) and locality that in such a case $a^\alpha(z)$ lies in the center of $\mathfrak{g}$.

COROLLARY 4.7. *Let $\mathfrak{g}$ be a regular formal distribution Lie superalgebra spanned by mutually local formal distributions $a^\alpha(z)$ ($\alpha \in I$). Then the OPE coefficients of the commutators $[a^\beta(z), a^\gamma(w)]$ ($\beta, \gamma \in I$) are finite $\mathbb{C}$-linear combinations of the formal distributions $a^\alpha(w)$ and their derivatives and some central formal distributions.*

PROOF. Consider the vertex algebra $\widetilde{V}^0(\mathfrak{g})$. Due to Remark 4.7, the representation of $\mathfrak{g}/\operatorname{center}(\mathfrak{g})$ in $\widetilde{V}^0(\mathfrak{g})$ is faithful. We have $a^\alpha(z) = Y(a^\alpha_{(-1)}|0\rangle, z)$, hence, by Theorem 4.6 we obtain:

$$[a^\alpha(z), a^\beta(w)] = \sum_{j=0}^{N-1} Y\left(a^\alpha_{(j)} a^\beta_{(-1)}|0\rangle, w\right) \partial_w^{(j)} \delta(z - w).$$

But, by (4.6.3), each vector $a^\alpha_{(j)} a^\beta_{(-1)}|0\rangle = \left[a^\alpha_{(j)}, a^\beta_{(-1)}\right]|0\rangle$ is a finite linear combination of vectors $a^\gamma_{(-i-1)}|0\rangle$ with $i \in \mathbb{Z}_+$. The corollary now follows from (4.4.6) for $b = |0\rangle$. $\qquad\square$

The vertex algebras $\widetilde{V}^\lambda(\mathfrak{g})$ are called *universal vertex algebras* associated to $\mathfrak{g}$.

Consider now the example of a current (resp. supercurrent) algebra $\hat{\mathfrak{g}}$ (resp. $\hat{\mathfrak{g}}_{\text{super}}$) associated to a Lie superalgebra $\mathfrak{g}$. This is a Lie superalgebra spanned by formal distributions $a(z)$ (resp. $a(z)$, $\bar{a}(z)$), $a \in \mathfrak{g}$, and K, with commutation relations given by (2.5.6) (resp. (2.5.6), (2.5.7a) and (2.5.7b)). Taking $T = -\partial_t$, it is immediate that (4.7.2) holds. Hence we may apply Theorem 4.7. We obviously have:

$$
(4.7.6) \qquad
\begin{aligned}
\hat{\mathfrak{g}}_{--} &= \mathfrak{g}[t] + \mathbb{C}K, & T\hat{\mathfrak{g}}_{--} &= \mathfrak{g}[t]; \\
(\hat{\mathfrak{g}}_{\text{super}})_{--} &= \mathfrak{g}[t, \theta] + \mathbb{C}K, & T(\hat{\mathfrak{g}}_{\text{super}})_{--} &= \mathfrak{g}[t, \theta].
\end{aligned}
$$

Thus, condition (4.7.3) gives us the following possibilities for λ:

$$\lambda(\mathfrak{g}[t]) \ (\text{resp. } \lambda(\mathfrak{g}[t, \theta])) = 0, \qquad \lambda(K) = k \in \mathbb{C}.$$

We shall denote the corresponding $\hat{\mathfrak{g}}$- (resp. $\hat{\mathfrak{g}}_{\text{super}}$-) module by $\widetilde{V}^k(\hat{\mathfrak{g}})$ (resp. $\widetilde{V}^k(\hat{\mathfrak{g}}_{\text{super}})$). By Theorem 4.7, $\widetilde{V}^k(\hat{\mathfrak{g}})$ and $\widetilde{V}^k(\hat{\mathfrak{g}}_{\text{super}})$ are vertex algebras, which are called *universal affine vertex algebras*.

In the special case when $\mathfrak{g}$ is a commutative Lie superalgebra, the universal affine vertex algebras are simple (i.e., have no non-zero ideals), provided that $k \neq 0$, due to Theorems 3.5 and 3.6. In this case the universal affine vertex algebra $\widetilde{V}^k(\hat{\mathfrak{g}})$ is called the *free bosonic vertex algebra* and is usually denoted by $B^k(\mathfrak{g})$.

One argues similarly in the case of the Clifford affinization C_A, defined by commutation relations (2.5.8). The corresponding vacuum vertex algebra (which is simple if $k \neq 0$ due to Theorem 3.6) is called the *free fermionic vertex algebra* and is usually denoted by $F^k(A)$. Note that for a commutative $\mathfrak{g}$ one has:

$$(4.7.7) \qquad \widetilde{V}^k(\hat{\mathfrak{g}})_{\text{super}} \simeq B^k(\mathfrak{g}) \otimes F^k(\bar{\mathfrak{g}}),$$

where the bar signifies the change of parity on $\mathfrak{g}$.

Let us demonstrate now on the example of currents $a(z)$, $a \in \mathfrak{g}$, how to use the "non-commutative" Wick formula. We shall work in the universal affine vertex algebra $\widetilde{V}^k(\mathfrak{g})$. By (2.5.6) we have

$$a(z)_\lambda b(z) = [a,b](z) + \lambda(a|b)k\,.$$

Hence by formula (3.3.12), we have

(4.7.8)

$$a(z)_\lambda : b(z)c(z) : =: [a,b](z)c(z) : +p(a,b) : b(z)[a,c](z) :$$

$$+ \lambda\left((a|b)kc(z) + p(a,b)(a|c)kb(z) + [[a,b],c]\,(z)\right) + \frac{\lambda^2}{2}k\,([a,b]|c)\,.$$

Thus, we obtain the following OPE:

$$a(z) : b(w)c(w) : \sim \frac{: [a,b](w)c(w) : +p(a,b) : b(w)[a,c](w) :}{z - w}$$

$$+ \frac{[[a,b],c]\,(w) + (a|b)kc(w) + p(a,b)(a|c)kb(w)}{(z-w)^2}$$

$$(4.7.9) \qquad\qquad + \frac{k\,([a,b]|c)}{(z-w)^3}\,.$$

4.8. Borcherds identity

THEOREM 4.8. *Let $F(z,w)$ be a rational function in z and w with poles only at $z = 0$, $w = 0$ or $z = w$. Then for any elements a and b of a vertex algebra V*

one has the following Borcherds identity:

(4.8.1)

$$\operatorname{Res}_{z-w} Y(Y(a, z-w)b, w)i_{w,z-w}F(z,w)$$
$$= \operatorname{Res}_z (Y(a,z)Y(b,w)i_{z,w}F(z,w) - p(a,b)Y(b,w)Y(a,z)i_{w,z}F(z,w)).$$

PROOF. It suffices to prove (4.8.1) for

$$F(z,w) = z^m(z-w)^n w^l, \quad m,n,l \in \mathbb{Z}.$$

Taking the residues for this F, (4.8.1) becomes the following identity multiplied by w^l:

$$(4.8.2) \qquad \sum_{j=0}^{\infty} \binom{m}{j} Y\left(a_{(n+j)}b, w\right) w^{m-j} = \sum_{j=0}^{\infty}(-1)^j \binom{n}{j} a_{(m+n-j)} Y(b,w)w^j$$
$$- p(a,b) \sum_{j=0}^{\infty}(-1)^{n+j} \binom{n}{j} Y(b,w)a_{(m+j)} w^{n-j},$$

which is Borcherds identity for $F = z^m(z-w)^n$. In particular, we see that Borcherds identity holds for $F(z,w)$ iff it holds for $w^l F(z,w)$, $l \in \mathbb{Z}$. It follows that it suffices to prove (4.8.2) in the following two cases:

$$\text{case 1:} \quad F = z^m, \ m \in \mathbb{Z}; \qquad \text{case 2:} \quad F = (z-w)^{-n-1}, \ n \in \mathbb{Z}_+.$$

But case 1 of (4.8.2) is precisely (4.6.4) and case 2 of (4.8.2) is precisely (4.4.6). $\square$

PROPOSITION 4.8. (a) *Borcherds identity is equivalent to the following three identities:*

(commutator) $\qquad\qquad [a_{(m)}, Y(b,z)] = \displaystyle\sum_{j=0}^{\infty} \binom{m}{j} Y\left(a_{(j)}b, z\right) z^{m-j},$

(normally ordered product) $\quad : Y(a,z)Y(b,z) : = Y\left(a_{(-1)}b, z\right)$

(derivative) $\qquad\qquad\qquad\qquad \partial Y(a,z) = Y(Ta, z).$

(b) *The following set of Borcherds axioms is an equivalent set of axioms of a vertex algebra:*

(partial vacuum) $|0\rangle_{(n)} a = \delta_{n,-1} a, \qquad a_{(-1)}|0\rangle = a;$

(4.8.3)

$$\text{(Borcherds identity)} \sum_{j=0}^{\infty} \binom{m}{j} \left(a_{(n+j)}b\right)_{(m+k-j)} c = \sum_{j=0}^{\infty} (-1)^j \binom{n}{j} a_{(m+n-j)}\left(b_{(k+j)}c\right)$$

$$- p(a,b) \sum_{j=0}^{\infty} (-1)^{j+n} \binom{n}{j} b_{(n+k-j)}\left(a_{(m+j)}c\right) (k,m,n \in \mathbb{Z}).$$

PROOF. (a) follows immediately from the proof of Theorem 4.8. Since (4.8.3) is an equivalent form of (4.8.2), our axioms listed in Section 1.3 imply Borcherds axioms (due to Theorem 4.8). Conversely, suppose that Borcherds axioms hold. Taking $b = |0\rangle$ and $F = 1$ in (4.8.1) we get $a_{(j)}|0\rangle = 0$ for $j \geq 0$, giving the vacuum axiom of Section 1.3. Letting $Ta = a_{(-2)}|0\rangle$, applying both sides of (4.8.3) to $|0\rangle$ and letting $m = 0$, $k = -2$ gives the translation covariance axiom. Finally, taking $F = z^m(z-w)^n$ for $n \gg 0$, we obtain the locality axiom from (4.8.1). $\qquad\square$

COROLLARY 4.8. *Borcherds identity holds for any three mutually local fields* $a(z)$, $b(z)$ *and* $c(z)$ *and any* $m,k,n \in \mathbb{Z}$:

(4.8.4)
$$\sum_{j=0}^{\infty} \binom{m}{j} \left(a(z)_{(n+j)}b(z)\right)_{(m+k-j)} c(z)$$

$$= \sum_{j=0}^{\infty} (-1)^j \binom{n}{j} \left(a(z)_{(m+n-j)} \left(b(z)_{(k+j)}c(z)\right)\right)$$

$$-(-1)^n p(a,b) b(z)_{(n+k-j)}\left(a(z)_{(m+j)}c(z)\right)) \ .$$

PROOF. Consider the (local) linear field algebra generated by the fields $a(z)$, $b(z)$ and $c(z)$. By Theorem 4.1, this is a vertex algebra, hence we may apply (4.8.3). $\qquad\square$

REMARK 4.8. (a) Letting $n = 0$ in (4.8.4), we get formula (3.3.7) for all $m,n \in \mathbb{Z}$, (not only for $m \in \mathbb{Z}_+$). Note, however, that (3.3.7) is a stronger statement in that respect that we do not make any assumptions on locality there.

(b) Recall that (4.8.4) for $m \in \mathbb{Z}_+$, $n = 0$ and $k = -1$ is the non-commutaive Wick formula (3.3.10). Another important special case of (4.8.4) is $m = 0$, $n = k =

-1, which is the "quasiassociativity" of the normally ordered product:

$$(4.8.5) \qquad :: a(z)b(z) : c(z) : - : a(z) : b(z)c(z) ::$$

$$= \sum_{j=0}^{\infty} a(z)_{(-j-2)}(b(z)_{(j)}c(z)) + p(a,b)b(z)_{(-j-2)}(a(z)_{(j)}c(z)).$$

EXAMPLE 4.8. Let $\alpha(z)$ be the free bosonic field (cf. Example 3.5) with the OPE

$$\alpha(z)\alpha(w) \sim \frac{1}{(z-w)^2}.$$

Then formula (4.8.5) gives:

$$:: \alpha(z)\alpha(z) : \alpha(z) : - : \alpha(z) : \alpha(z)\alpha(z) ::= \partial_z^2 \alpha(z),$$

i.e. associativity of the normally ordered product fails even for free fields.

4.9. Graded and Möbius conformal vertex algebras

A vertex algebra V is called *graded* if there is an even diagonalizable operator H on V such that

$$(4.9.1) \qquad [H, Y(a,z)] = z\partial Y(a,z) + Y(Ha,z).$$

Note that (4.9.1) means that the field $Y(a,z)$ has conformal weight $\Delta \in \mathbb{C}$ with respect to the Hamiltonian $\mathrm{ad}H$ (see Definition 2.6a) iff $Ha = \Delta a$. By abuse of terminology, we shall call H a Hamiltonian of a vertex algebra V if (4.9.1) holds.

As in Section 2.6, writing the field of conformal weight Δ in the form

$$Y(a,z) = \sum_{n \in -\Delta + \mathbb{Z}} a_n z^{-n-\Delta},$$

so that

$$(4.9.2) \qquad a_{(n)} = a_{n-\Delta+1},$$

we see that (4.9.1) is equivalent to

$$(4.9.3) \qquad [H, a_n] = -na_n.$$

Note that (1.3.4) becomes:

$$(4.9.4) \qquad [T, a_n] = (-n - \Delta + 1)a_{n-1},$$

and (4.9.1) for $a = |0\rangle$ gives

$$(4.9.5) \qquad\qquad\qquad\qquad H|0\rangle = 0.$$

It follows that

$$(4.9.6) \qquad\qquad\qquad\qquad [H, T] = T$$

since both sides commute in the same way with all a_n and both annihilate $|0\rangle$.

Consider the eigenspace decomposition of V with respect to H:

$$V = \bigoplus_j V^{(j)}.$$

Note that, by (4.9.3) and (4.9.4) one has:

$$(4.9.7) \qquad\qquad a_n V^{(j)} \subset V^{(j-n)}, \quad T V^{(j)} \subset V^{(j+1)}.$$

It is clear that a graded vertex algebra has a unique maximal graded ideal and that the corresponding quotient vertex algebra is simple.

REMARK 4.9a. If V is a graded vertex algebra, one usually considers the "restricted" dual space:

$$V^* = \bigoplus_j V^{(j)*}$$

and the matrix coefficients of fields or their products, like

$$M^{a,b}_{v^*,v}(z, w) = \langle v^*, Y(a, z) Y(b, w) v \rangle, \quad v \in V^{(i)}, \quad v^* \in V^{(j)*}.$$

Then provided that the real part of the spectrum of H is bounded below this matrix coefficient converges to a rational function in the domain $|z| > |w|$, and we may extend it analytically to the domain $z \neq 0$, $w \neq 0$, $z \neq w$. Then the equality of all matrix coefficients is equivalent to the equality of the product of fields. For example, the locality is equivalent to the equality of all rational functions:

$$M^{a,b}_{v^*,v}(z, w) = p(a, b) M^{b,a}_{v^*,v}(w, z).$$

In this approach the proofs are somewhat simpler (for example, Theorem 4.6 is then immediate by Goddard's Uniqueness theorem) and (4.6.1a and b) makes sense (as an equality of the matrix coefficients). However, this approach is restricted to the graded case only.

Using (4.9.2) we rewrite (4.6.3) and (4.6.4) in a graded form ($m, n \in \mathbb{Z}$):

$$(4.9.8) \qquad [a_m, b_n] = \sum_{j \in \mathbb{Z}_+} \binom{m + \Delta - 1}{j} (a_{j - \Delta + 1} b)_{m+n},$$

$$(4.9.9) \qquad [a_m, Y(b, z)] = \sum_{j \in \mathbb{Z}_+} \binom{m + \Delta - 1}{j} Y\left(a_{j - \Delta + 1} b, z\right) z^{m - j + \Delta - 1}.$$

Hence $\operatorname{Lie} V$ becomes a $\mathbb{Z}$-graded Lie algebra, the gradation being given by the eigenvalues of $\operatorname{ad} H$. Note that $\operatorname{ad} T$ is a derivation of $\operatorname{Lie} V$ that shifts this gradation by -1.

The following remark allows one to construct Hamiltonians.

REMARK 4.9b. Let V be a vertex algebra and let H be a diagonalizable operator on the space V such that $H|0\rangle = 0$. Suppose that V is strongly generated by a collection of fields $Y(a^\alpha, z)$ such that

$$[H, Y(a^\alpha, z)] = (z\partial + \Delta_\alpha) Y(a^\alpha, z), \quad \Delta_\alpha \in \mathbb{C}.$$

Then H is a Hamiltonian of the vertex algebra V. This follows from (4.9.4) and (4.9.5).

The following remark allows one to construct automorphisms of a vertex algebra.

REMARK 4.9c. Suppose that $V = \bigoplus_j V^{(j)}$ is a graded vertex algebra such that $\dim V^{(j)} < \infty$ for all j, and let $a \in V_0^{(1)}$. It follows from (4.9.8) that a_0 is a derivation of the vertex algebra V which preserve the gradation. Hence the series e^{a_0} converges to an automorphism of the vertex algebra V called an *inner automorphism* of V.

The following proposition allows one to compute the vacuum subalgebra of a vertex algebra.

PROPOSITION 4.9. *Let V be a vertex algebra graded by a Hamiltonian H all of whose eigenvalues are non-negative. Suppose that there exists an operator T^* on V such that*

$$(4.9.10) \qquad [H, T^*] = -T^*, \qquad [T^*, T] = 2H.$$

Then

(a) $\operatorname{Vac} V \subset V^{(0)}\,(=\operatorname{Ker} H)$.

(b) *The representation of the Lie algebra* $\mathfrak{r} = \mathbb{C}T + \mathbb{C}H + \mathbb{C}T^*$ *on V is completely reducible if and only if* $\operatorname{Vac} V = V^{(0)}$ *and* $\operatorname{Vac} V \cap \operatorname{Im} T^* = 0$.

PROOF. Due to (4.9.6) and (4.9.10), $\mathfrak{r}$ is a Lie algebra isomorphic to $sl_2(\mathbb{C})$. It is clear that every irreducible subquotient of the $\mathfrak{r}$-module V either is a 1-dimensional $\mathfrak{r}$-module or is a Verma module with respect to the Borel subalgebra $\mathbb{C}H + \mathbb{C}T^*$ with negative highest weight. Proposition now follows from the elementary highest weight representation theory of $sl_2(\mathbb{C})$ (or one can apply the general Proposition 9.9 from [**K2**]). $\square$

EXAMPLE 4.9a. Let $\mathfrak{g}$ be a regular Lie superalgebra of formal distributions. Suppose that $\mathfrak{g}$ is graded with the Hamiltonian H (see (3.4.1)). Then $H\mathfrak{g}_{--} \subset \mathfrak{g}_{--}$ (due to (4.9.6)) and hence, due to Remark 4.9a, the associated vertex algebras $\widetilde{V}^\lambda(\mathfrak{g})$ are graded. The simple quotient vertex algebra of $\widetilde{V}^\lambda(\mathfrak{g})$ by the maximal graded ideal is denoted by $V^\lambda(\mathfrak{g})$. Furthermore, suppose that there exists a derivation T^* of $\mathfrak{g}$ such that (4.9.10) holds. Then $T^*\mathfrak{g}_{--} \subset \mathfrak{g}_{--}$, hence we get an induced operator on $\widetilde{V}^\lambda(\mathfrak{g})$ which we again denote by T^*. It is easy to see that the maximal graded ideal of $\widetilde{V}^\lambda(\mathfrak{g})$ is T^*-invariant, hence we get an induced operator T^* on the vertex algebra $V^\lambda(\mathfrak{g})$ which still satisfies (4.9.10).

The following is a special case of Example 4.9a.

EXAMPLE 4.9b. It follows from Example 2.6 that the universal affine vertex algebras $\widetilde{V}^k(\widehat{\mathfrak{g}})$ and $\widetilde{V}^k(\widehat{\mathfrak{g}}_{\mathrm{super}})$ are graded, the conformal weights of currents (resp. supercurrents) being 1 (resp. 1/2). In particular, the free bosonic (resp. free fermionic) vertex algebra is graded by taking the conformal weight of free generating bosons (resp. fermions) equal 1 (resp. 1/2). The simple graded quotient of the universal affine vertex algebra $\widetilde{V}^k(\widehat{\mathfrak{g}})$ (resp. $\widetilde{V}^k(\widehat{\mathfrak{g}}_{\mathrm{super}})$) is called an *affine* (resp. *superaffine*) *vertex algebra* and is denoted by $V^k(\widehat{\mathfrak{g}})$ (resp. $V^k(\widehat{\mathfrak{g}}_{\mathrm{super}})$). Note that $T^* = -t^2\partial_t$ (resp. $-t^2\partial_t - \frac{1}{2}t\theta\partial_\theta$) is a derivation of the algebra of currents $\widehat{\mathfrak{g}}$ (resp. supercurrents $\widehat{\mathfrak{g}}_{\mathrm{super}}$). Thus, the vertex algebras $V^k(\widehat{\mathfrak{g}})$ and $V^k(\widehat{\mathfrak{g}}_{\mathrm{super}})$ satisfy the conditions of Proposition 4.9. Since $\operatorname{Ker} H = \mathbb{C}|0\rangle$ we obtain that in both cases the vacuum subalgebra is $\mathbb{C}|0\rangle$.

The following definition is motivated by (1.2.6c) and the subsequent discussion.

DEFINITION 4.9. A graded by H vertex algebra V is called *Möbius*-conformal if there exists an operator T^* on V which decreases the conformal weight by 1 and such that for any $a \in V$ one has:

$$(4.9.11) \qquad [T^*, Y(a, z)] = z^2 \partial Y(a, z) + 2z Y(Ha, z) + Y(T^*a, z).$$

Letting $a = |0\rangle$ in (4.9.11), we get from (4.9.5) and the axioms of V:

$$(4.9.12) \qquad\qquad T^*|0\rangle = 0.$$

We also have:

$$(4.9.13) \qquad [T^*, a_n] = -(n - \Delta + 1)a_{n+1} + (T^*a)_{n+1}.$$

Combining (4.9.12),(4.9.13) and (4.9.3),(4.9.4), we see that (4.9.10) is satisfied.

Motivated by (1.2.5c), a field $Y(a, z)$ of conformal weight Δ of a Möbius-conformal vertex algebra is called *quasiprimary* if

$$[T^*, Y(a, z)] = (z^2 \partial + 2\Delta z) Y(a, z).$$

REMARK 4.9d. $Y(a, z)$ is a quasiprimary field of conformal dimension Δ iff

$$(4.9.14) \qquad\qquad Ha = \Delta a, \qquad T^*a = 0.$$

Note that if the representation of the Lie algebra $\mathbb{C}T + \mathbb{C}H + \mathbb{C}T^*$ in V is completely reducible, the vectors $T^{*n}a$, where a satisfies (4.9.14) for some Δ and $n \in \mathbb{Z}_+$, span V. (Due to Proposition 4.9, this condition holds if all eigenvectors, except for the $\mathbb{C}|0\rangle$, of L_0 have positive eigenvalues.) Hence in this case the quasiprimary fields along with all their derivatives span the space of all fields of the vertex algebra V.

Recall that the axiom of translation covariance integrates to the following equality of formal distributions in z and λ in the domain $|\lambda| < |z|$:

$$(4.9.15) \qquad\qquad e^{\lambda T} Y(a, z) e^{-\lambda T} = Y(a, z + \lambda).$$

Similarly, relation (4.9.1) integrates to (cf. Section 1.2):

$$(4.9.16) \qquad\qquad \lambda^H Y(a, z) \lambda^{-H} = Y(\lambda^H a, \lambda z).$$

Indeed, (4.9.1) is equivalent to (4.9.3) which integrates to $\lambda^H a_n \lambda^{-H} = \lambda^{-n} a_n$, which is equivalent to (4.9.16).

Finally, (4.9.11) integrates to the following covariance relation in the domain $|\lambda z| < 1$:

$$(4.9.17) \qquad e^{\lambda T^*} Y(a, z) e^{-\lambda T^*} = Y\left(e^{\lambda(1-\lambda z)T^*} (1 - \lambda z)^{-2H} a, \frac{z}{1 - \lambda z} \right).$$

(Note that this reduces to a special case of (1.2.4) if a satisfies (4.9.14).) In order to prove this relation, write:

$$e^{\lambda T^*} Y(a, z) e^{-\lambda T^*} = Y\left(A(\lambda)a, \frac{z}{1 - \lambda z} \right),$$

where $A(\lambda)$ is a formal power series in λ with coefficients in $\operatorname{Hom}(V, \operatorname{End}V\left[\left[z, z^{-1}\right]\right])$ and constant term I. Differentiating both sides by λ and using (4.9.11) we obtain an equation on $A(\lambda)$:

$$\frac{dA(\lambda)}{d\lambda} = z^2 \partial_z A(\lambda) + 2z A(\lambda) H + A(\lambda) T^*$$

which has a unique solution with constant term I (by Lemma 4.1). To check that $A(\lambda) = e^{\lambda(1-\lambda z)T^*} (1 - \lambda z)^{-2H}$ is a solution to this equation, we use (4.9.10) (and that $\operatorname{ad}T^*$ is a derivation).

In what follows, we shall perform calculations in the Lie algebra $sl_2\left(\mathbb{C}((\lambda))\right)$ and the corresponding group $SL_2\left(\mathbb{C}((\lambda))\right)$, where $\mathbb{C}((\lambda))$ stands for the field of Laurent series in λ, which act on $\mathbb{C}((\lambda)) \otimes_{\mathbb{C}} V$ via the identification

$$T = \begin{pmatrix} 0 & 1 \\ 0 & 0 \end{pmatrix}, \quad 2H = \begin{pmatrix} 1 & 0 \\ 0 & -1 \end{pmatrix}, \quad T^* = \begin{pmatrix} 0 & 0 \\ -1 & 0 \end{pmatrix}.$$

(In the previous calculations we, in fact, kept this identification in mind.) Formulas (4.9.15) and (4.9.17) give us respectively:

(4.9.18)
$$\begin{pmatrix} 1 & \lambda \\ 0 & 1 \end{pmatrix} Y(a, z) \begin{pmatrix} 1 & -\lambda \\ 0 & 1 \end{pmatrix} = Y(a, z + \lambda) \quad (|\lambda| < |z|)$$

(4.9.19)
$$\begin{pmatrix} 1 & 0 \\ -\lambda & 1 \end{pmatrix} Y(a, z) \begin{pmatrix} 1 & 0 \\ \lambda & 1 \end{pmatrix} = Y\left(e^{\lambda(1-\lambda z)T^*} (1 - \lambda z)^{-2H} a, \frac{z}{1 - \lambda z} \right) \quad (|\lambda z| < 1).$$

Using that

$$\begin{pmatrix} 0 & \lambda \\ -\lambda^{-1} & 0 \end{pmatrix} = \begin{pmatrix} 1 & \lambda \\ 0 & 1 \end{pmatrix} \begin{pmatrix} 1 & 0 \\ -\lambda^{-1} & 1 \end{pmatrix} \begin{pmatrix} 1 & \lambda \\ 0 & 1 \end{pmatrix},$$

we deduce from (4.9.18) and (4.9.19) (cf. [**B1**] and [**DGM**]):

(4.9.20)
$$\begin{pmatrix} 0 & \lambda \\ -\lambda^{-1} & 0 \end{pmatrix} Y(a, z) \begin{pmatrix} 0 & -\lambda \\ \lambda^{-1} & 0 \end{pmatrix} = Y\left(e^{-\lambda^{-2}zT^*}(-\lambda^{-1}z)^{-2H}a, -\frac{\lambda^2}{z}\right).$$

This formula holds over $\mathbb{C}((\lambda))$ provided that T^* is locally nilpotent on V (which is the case when the spectrum of H is bounded below).

4.10. Conformal vertex algebras

It is well-known that even locally the only orientation preserving conformal transformations of the Minkowski space of dimension $d > 2$ are in the conformal group described in Section 1.1. Of course in the $d = 2$ case the situation is dramatically different—any transformation of the form $t \mapsto f(t)$, $\bar{t} \mapsto f(\bar{t})$, where $(t, \bar{t})$ are light-cone coordinates and f is a smooth function with a non-vanishing derivative, is conformal. For that reason, the term "conformal" 2-dimensional QFT is reserved for the case when covariance holds for this much larger group. We give now the precise definition, which is motivated by the notion of the energy-momentum field of a QFT. Recall that a field $L(z)$ with the OPE (2.6.5) in which $C = cI$, $c \in \mathbb{C}$, is called a Virasoro field with central charge c.

DEFINITION 4.10. [**B1**] A *conformal vector* of a vertex algebra V is an even vector ν such that the corresponding field $Y(\nu, z) = \sum_{n \in \mathbb{Z}} L_n z^{-n-2}$ is a Virasoro field with central charge c which has the following properties:

(a) $L_{-1} = T$,

(b) L_0 is diagonalizable on V.

The number c is called the *central charge* of ν. A vertex algebra endowed with a conformal vector ν is called a *conformal* vertex algebra of *rank* c. The field $Y(\nu, z)$ is called an *energy-momentum field* of the vertex algebra V.

THEOREM 4.10. (a) *Suppose a vector* $\nu' \in V$ *satisfies properties*

    (i) $L_{-1} = T$,

    (ii) $L_2\nu = \frac{c}{2}|0\rangle$ *for some* $c \in \mathbb{C}$,

    (iii) $L_n\nu = 0$ *for* $n > 2$.

Then there exists $a \in \mathrm{Vac}\, V$ *such that* $\nu = \nu' - a$ *satisfies* (i)–(iii) *and*

(iv) $L_0\nu = 2\nu$.

(b) *If (i)–(iv) hold, then $Y(\nu, z)$ is a Virasoro field with central charge c.*

(c) *A vector $\nu \in V$ is a conformal vector iff it satisfies (i)–(iv) and*

(v) *L_0 is diagonalizable on V.*

(d) *If (i) and (v) hold, then V is a graded vertex algebra with respect to L_0.*

(e) *A conformal vertex algebra is Möbius-conformal with $H = L_0$ and $T^* = L_1$.*

PROOF. Due to (4.6.2a) we have the following OPE:

$$Y(\nu', z)Y(\nu', w) \sim \sum_{n \geq -1} \frac{Y(L_n\nu', w)}{(z - w)^{n+2}},$$

hence using (i)–(iii) and (4.4.7) we obtain:

$$(4.10.1)\quad Y(\nu', z)Y(\nu', w) \sim \frac{c/2}{(z - w)^4} + \frac{Y(L_1\nu', w)}{(z - w)^3} + \frac{Y(L_0\nu', w)}{(z - w)^2} + \frac{\partial Y(\nu', w)}{z - w}.$$

It follows from Theorem 2.6(a) that

$$(4.10.2)\qquad\qquad\qquad L_1\nu' = 0,$$

$$\partial Y\left(\nu' - \tfrac{1}{2}L_0\nu', w\right) = 0.$$

Hence $\nu' - \tfrac{1}{2}L_0\nu' = a \in \mathrm{Vac}\, V$. Due to Remark 4.4a, $L_n a = 0$ for $n \geq -1$ and $a_{(k)} = \delta_{k,-1}a_{(k)}$. Hence $\nu = \nu' - a$ satisfies (i)–(iv), proving (a). Formula (4.10.1) along with (4.10.2) proves (b). (c) follows from (b) and the OPE for the Virasoro field. By (4.6.2a) and (4.9.9) we have for any $a \in V$:

$$(4.10.3)\qquad Y(\nu, z)Y(a, w) \quad \sim \quad \sum_{n \geq -1} \frac{Y(L_n a, w)}{(z - w)^{n+2}},$$

$$(4.10.4)\qquad [L_m, Y(a, z)] \quad = \quad \sum_{j \geq -1} \binom{m + 1}{j + 1} Y(L_j a, z) z^{m-j}.$$

Equation (4.10.4) for $m = 0$ proves (d). (e) follows from (c).   □

REMARK 4.10. Let $\nu \in V$ be such that $Y(\nu, z)$ is a Virasoro field and $\nu_{(1)}$ is a diagonalizable operator on V. Then the subspace $\{a \in V\,|\,[\nu_{(0)}, Y(a, z)] = \partial Y(a, z)\}$ is the maximal subalgebra of V for which ν is a conformal vector.

If $L_0 a = \Delta a$, we have by (4.10.3) and (4.4.7)

$$Y(\nu, z)Y(a, w) \sim \frac{\partial Y(a, w)}{z - w} + \frac{\Delta Y(a, w)}{(z - w)^2} + \cdots.$$

A field $Y(a, z)$ of a conformal vertex algebra V is called *primary* of conformal weight Δ if there are no extra terms in the above OPE:

$$Y(\nu, z)Y(a, w) \sim \frac{\partial Y(a, w)}{z - w} + \frac{\Delta Y(a, w)}{(z - w)^2} .$$

COROLLARY 4.10. *The field $Y(a, z)$ is primary of conformal weight Δ iff one of the following equivalent conditions hold:*

(i) $L_n a = \delta_{n,0} \Delta a$ *for all* $n \in \mathbb{Z}_+$;

(ii) $[L_m, Y(a, z)] = z^m(z\partial + \Delta(m + 1))Y(a, z)$, $m \in \mathbb{Z}$;

(iii) $[L_m, a_n] = ((\Delta - 1)m - n)a_{m+n}$, $m, n \in \mathbb{Z}$.

Note that a primary field is always quasiprimary.

We consider now some examples.

PROPOSITION 4.10. (a) *Let $\mathfrak{h}$ be a finite-dimensional superspace with a super-symmetric non-degenerate bilinear form, let $\{a^i\}$ and $\{b^i\}$ be dual bases of $\mathfrak{h}$, let $b \in \mathfrak{h}_{\bar{0}}$ and let k be a non-zero complex number. Then*

$$\nu(b) := \frac{1}{2k} \sum_i a^i_{(-1)} b^i_{(-1)} |0\rangle + b_{(-2)} |0\rangle$$

*is a conformal vector of the vertex algebra $B^k(\mathfrak{h})$ with central charge $c = $ *sdim $\mathfrak{h} - 12(b \mid b)k$.

(b) *Let A be a finite-dimensional superspace with a skew supersymmetric non-degenerate bilinear form, let $\{\varphi^i\}$ and $\{\psi^i\}$ be dual bases of A and let k be a non-zero complex number. Then*

$$\nu := \frac{1}{2k} \sum_i \varphi^i_{(-\frac{3}{2})} \psi^i_{(-\frac{1}{2})} |0\rangle$$

is a conformal vector of the vertex algebra $F^k(A)$ with central charge $c = -\frac{1}{2}$ sdim A.

PROOF. Note that by (4.4.6) and (4.4.7) we have (see Section 3.5):

$$(4.10.5) \qquad Y(\nu(b), z) = L^b(z) = \sum_n L^b_n z^{-n-2}.$$

Recall from Section 3.5 that $L^b(z)$ is a Virasoro field with c given above, and for any $a \in \mathfrak{h}$ we have:

$$(4.10.6) \qquad [L^b_{-1}, a(z)] = \partial a(z),$$

$$(4.10.7) \qquad L^b_0 = H.$$

Now property (a) of $\nu(b)$ follows from (4.10.6), and property (b) from (4.10.7), proving (a). In a similar way, (b) follows from the discussion in Section 3.6. $\square$

Note that in the case of the vertex algebra $B^k(\mathfrak{h})$, all free bosons $a(z)$ have conformal weight 1 with respect to L_0^b, but they are primary iff $b = 0$ (see (3.5.12)). In the case of $F^k(A)$, all free fermions are primary fields of conformal weight $\frac{1}{2}$ (see (3.6.4)).

One can construct in a similar way the conformal vector for an arbitrary vacuum affine vertex algebra $\tilde{V}^k(\mathfrak{g})$ or $\tilde{V}^k(\mathfrak{g}_{\mathrm{super}})$ (under a suitable assumption on k) but the calculation is somewhat more involved and will be done later (see Section 5.7).

EXAMPLE 4.10. The Virasoro algebra Vir (defined by commutation relations (2.6.6)) is spanned by formal distribution $L(z) = \sum_{n \in \mathbb{Z}} L_n z^{-n-2}$ and C (the OPE being given by (2.6.5)). The derivation $T = \mathrm{ad}L_{-1}$ satisfies (4.7.1), and $H = \mathrm{ad}L_0$ is a Hamiltonian with respect to which $L(z)$ and C have conformal weights 2 and 0 respectively. Note that

$$\mathrm{Vir}_{--} = \mathbb{C}C + \sum_{n \geq -1} \mathbb{C}L_n, \quad T(\mathrm{Vir}_{--}) = \sum_{n \geq -1} \mathbb{C}L_n.$$

Hence (due to Theorem 4.7) the associated universal vertex algebras $\tilde{V}^c(\mathrm{Vir})$ are parametrized by a complex number c $(= \lambda(C))$. All these vertex algebras (and their quotients) are conformal with the conformal vector

$$\nu = L_{-2}|0\rangle,$$

so that $Y(\nu, z) = L(z)$. In particular these vertex algebras are graded with the Hamiltonian L_0, and we have the corresponding simple conformal vertex algebras $V^c(\mathrm{Vir})$ (of rank c), called the *Virasoro vertex algebras*.

The vertex algebras $V^c(\mathrm{Vir})$ are characterized by the property of being simple graded vertex algebras strongly generated by a non-free field of conformal weight 2. Indeed, writing this field in the form $L(z) = \sum_n L_n z^{-n-2}$, we have:

$$V^{(j)} = \mathbb{C}\delta_{j0}|0\rangle \text{ if } j \leq 1, \qquad V^{(2)} = \mathbb{C}\nu, \text{ where } \nu = L_{-2}|0\rangle,$$

so that $L(z) = Y(\nu, z)$. Hence we have:

$$L(z)L(w) = \frac{c}{(z-w)^4} + \frac{2\alpha L(w)}{(z-w)^2} + \frac{\Psi(w)}{z-w},$$

for some $c, \alpha \in \mathbb{C}$ and some field $\Psi(w)$. By Theorem 2.6 we conclude that $\Psi(w) = \alpha \partial L(w)$, and hence $\alpha \neq 0$ since $L(z)$ is a non-free field. We can rescale ν so that $L(z)$ becomes a Virasoro field.

Note that holomorphic vertex algebras do not admit a conformal structure since the Virasoro field is not holomorphic.

4.11. Field algebras

Field algebras generalize vertex algebras in the same way as unital associative algebras generalize unital commutative associative algebras.

A *field algebra* V is defined by the same data as a vertex algebra, but weaker axioms (cf. Proposition 4.8(b)):

(partial vacuum): $Y(|0\rangle, z) = I_V$, $a_{(-1)}|0\rangle = a$,

(n-th product): $Y(a_{(n)}b, z) = Y(a, z)_{(n)}Y(b, z)$, $n \in \mathbb{Z}$.

Note that the n-th product axiom is nothing else but Borcherds identity in the form (4.8.1) for $F = (z-w)^n$. As in the proof of Theorem 4.8, it follows that (4.8.1) holds for $F = z^m$ with $m \in \mathbb{Z}_+$. Hence the n-th product axiom implies (4.6.4) for $m \in \mathbb{Z}_+$, and in particular, the axiom (C3) of conformal algebra.

As in the case of vertex algebra, the axioms of a field algebra imply:

$$(4.11.1) \qquad Y(a, z)|0\rangle|_{z=0} = a, \quad Y(|0\rangle, z) = I_V,$$

$$(4.11.2) \qquad Y(Ta, z) = \partial Y(a, z) = [T, Y(a, z)],$$

where $T \in \mathrm{End} V$ is defined by $Ta = a_{(-2)}|0\rangle$. The n-th product axiom for $n >> 0$ implies *weak locality*:

$$(4.11.3) \qquad \mathrm{Res}_z (z - w)^N [Y(a, z), Y(b, w)] = 0 \text{ for } N >> 0.$$

Note that weak locality of fields $a(z)$ and $b(z)$ means that $a(z)_{(n)}b(z) = 0$ for $n \geq N$, some N. (Unlike the usual locality, this is not a symmetric property.) Then, clearly, $(za(z))_{(n)}b(z) = 0$ for $n \geq N$. Using this remark, one can extend the proof of Dong's lemma to the weakly local case (assuming that all ordered pairs are weakly local).

EXAMPLE 4.11. Recall that any two local fields satisfy the skewsymmetry relation (3.3.6). This, however, fails for weakly local fields. In order to construct a counterexample, consider the free bosonic field $\alpha(z) = \sum_{n \in \mathbb{Z}} \alpha_n z^{-n-1}$ (cf. Example 3.5), and let $\beta(z) = \sum_{n>0} n^{-1} \alpha_n z^{-n}$. Then we have:

$$[\alpha(z), \beta(w)] = i_{w,z}(z - w)^{-1} .$$

Hence for $j \in \mathbb{Z}_+$ we have:

$$\alpha(z)_{(j)}\beta(z) = 0 , \quad \beta(z)_{(j)}\alpha(z) = \delta_{j0} .$$

Therefore both pairs (α, β) and (β, α) are weakly local, but (3.3.6) fails for $a = \alpha$, $b = \beta$, $n = 0$.

Recall that the -1st product axiom means:

$$(4.11.4) \qquad\qquad Y(a_{(-1)}b, z) =: Y(a, z)Y(b, z) : .$$

Replacing a by $T^n a$ and using (4.11.2), we see that (4.11.4) implies the n-th product axiom for $n < 0$.

Multiplying both sides of the n-th product axiom by $(-w)^{-n-1}$ and taking summation over $n \in \mathbb{Z}$, we obtain its equivalent form in the domain $|z| > |w|$:

$$(4.11.5) \qquad Y(Y(a, z)b, -w)c = Y(a, z - w)Y(b, -w)c$$

$$-p(a, b)Y(b, -w) \sum_{j \geq 0} \partial_w^j \delta(z - w) \operatorname{Res}_x x^{(j)} Y(a, x)c .$$

This is immediate by the following special case of Taylor's formula in the domain $|z| > |w|$:

$$i_{w,x}\delta((w + x) - z) = \sum_{j \geq 0} x^{(j)} \partial_w^j \delta(z - w) .$$

Formula (4.11.5) implies the *associativity* property in the domain $|z| > |w|$:

$$(4.11.6) \quad (z - w)^N Y(Y(a, z)b, -w)c = (z - w)^N Y(a, z - w)Y(b, -w)c \text{ for } N \gg 0 .$$

As in Section 1.4, it is easy to show that all holomorphic field algebras are obtained by taking a unital associative algebra V and its derivation T, and letting

$$Y(a, z)b = e^{zT}(a)b, \quad a, b \in V.$$

The general linear field algebra $g\ell f(U)$ defined in Section 3.2 is not a field algebra since the field property

$$(4.11.7) \qquad\qquad a_{(n)}b = 0 \quad \text{for} \quad n \gg 0$$

fails in general. However, if we take a collection of mutually weakly local fields $\{a^\alpha(z)\} \subset g\ell f(U)$, they generate a linear field algebra which is a field algebra. The n-th product axiom for $n \geq 0$ is implied by (3.3.7). Next, it is immediate to check (4.11.1) and (4.11.2). Weak locality is proved in the same way as Proposition 3.2. The n-th product axiom for $n < 0$ follows from (4.11.4) as explained above. Finally, the -1st product axiom is checked by a direct calculation.

We have the following field algebra analogs of the uniqueness and existence theorems (obtained jointly with Bojko Bakalov).

THEOREM 4.11. (a) *Let V be a field algebra. For each field $Y(a, z)$ define the "opposite" field $X(a, z)$ by the formula (cf. (4.2.1)):*

$$(4.11.8) \qquad X(a, z)b = p(a, b)e^{zT}Y(b, -z)a.$$

Let $B(z)$ be a field which is mutually local with all fields $X(a, z), a \in V$, on any $v \in V$, i.e.,

$$(z - w)^N[B(z), X(a, z)]v = 0 \text{ for } N \gg 0.$$

Suppose that (4.4.1) holds for some $b \in V$. Then $B(z) = Y(b, z)$.

(b) *Let V be a vector superspace, let $|0\rangle$ be an even vector and T an even endomorphism of V. Let $\{a^\alpha(z)\}_{\alpha \in A}$ and $\{b^\beta(z)\}_{\beta \in B}$ (A, B index sets) be two collections of fields such that each of them satisfies conditions (i)–(v) of Theorem 4.5 except that in (iii) "local" is replaced by "weakly local". Suppose, in addition, that all pairs $(a^\alpha(z), b^\beta(z))$ are local on any $v \in V$. Then formula (4.5.1) defines a unique structure of a field algebra on V such that $|0\rangle$ is the vacuum vector, T is the infinitesimal translation operator and (4.5.2) holds. The same conclusion holds if the family $\{a^\alpha(z)\}$ is replaced by the family $\{b^\beta(z)\}$ and the fields Y are replaced by the fields X. The two field algebra structures on V are related by (4.11.8).*

PROOF. It is similar to that of Theorems 4.4 and 4.5 using the observation that the associativity property (4.11.6) is equivalent to the locality of the pair $(Y(a, z), X(c, z))$ on $b \in V$.

$\square$

Taking in this theorem all fields of a field algebra, we obtain the following corollary.

COROLLARY 4.11. (a) *Vacuum and translation covariance axioms along with weak locality (4.11.3) and associativity (4.11.6) form an equivalent system of axioms of a field algebra.*

(b) *If $(V, |0\rangle, T, Y(a, z))$ is a field algebra, then $(V, |0\rangle, T, X(a, z))$, where $X(a, z)$ are defined by (4.11.8), is a field algebra as well.*

REMARK 4.11a. It follows from the above discussion that a field algebra with n-th products for $n \in \mathbb{Z}_+$ and $\partial = T$ satisfies all axioms of a conformal algebra, except the skewsymmetry axiom (C2), which may fail in view of Example 4.11.

REMARK 4.11b. It follows from the proof of Proposition 3.3(b) that two weakly local fields $a(z)$ and $b(z)$ for which the skewsymmetry property (3.3.6) holds, are local. Hence a field algebra satisfying the skewsymmetry property (4.2.2) is a vertex algebra. This follows also from Corollary 4.11.

CHAPTER 5

Examples of vertex algebras and their applications

5.1. Charged free fermions and triple product identity

Recall (see Section 2.5) that "charged free fermions" is a formal distribution Lie superalgebra spanned by coefficients of odd formal distributions $\psi^+(z)$ and $\psi^-(z)$ and an even (constant) formal distribution 1 commuting with $\psi^\pm(z)$ with the OPE (2.5.13). We shall denote this Lie superalgebra by C_{char}. It has a basis consisting of odd elements $\psi^\pm_{(n)}$ ($n \in \mathbb{Z}$) and an even central element 1 with commutation relations:

$$\left[\psi^+_{(m)}, \psi^-_{(n)}\right] = \delta_{m,-n-1}, \quad \left[\psi^\pm_{(m)}, \psi^\pm_{(n)}\right] = 0.$$

It is a regular formal distribution Lie superalgebra since it admits a derivation T defined by (4.7.1) (i.e., $T\psi^\pm_{(n)} = -n\psi^\pm_{(n-1)}$, $T1 = 0$).

Recall that, by Theorem 3.6, the Lie superalgebra C_{char} has a unique irreducible module, which we shall denote by F, such that the central element 1 is represented by the identity operator and there exists a non-zero vector $|0\rangle$ such that

$$\psi^\pm(z)_-|0\rangle = 0.$$

Due to Theorem 4.7, F is a (simple) vertex algebra with the vacuum vector $|0\rangle$ and generated by the fields $\psi^+(z)$ and $\psi^-(z)$.

The vertex algebra F has a 1-parameter family of conformal vectors ($\lambda \in \mathbb{C}$):

$$(5.1.1) \qquad \nu^\lambda = (1-\lambda)\psi^+_{(-2)}\psi^-_{(-1)}|0\rangle + \lambda\psi^-_{(-2)}\psi^+_{(-1)}|0\rangle.$$

Indeed, $Y\left(\nu^\lambda, z\right) = L^\lambda(z)$ (which is given by (3.6.14)), and by (3.6.15) we have:

$$(5.1.2) \qquad \begin{aligned} Y\left(\nu^\lambda, z\right)\psi^+(w) &\sim \frac{\partial\psi^+(w)}{z-w} + \frac{\lambda\psi^+(w)}{(z-w)^2}, \\ Y\left(\nu^\lambda, z\right)\psi^-(w) &\sim \frac{\partial\psi^-(w)}{z-w} + \frac{(1-\lambda)\psi^-(w)}{(z-w)^2}. \end{aligned}$$

It follows that $L^\lambda_{-1} = T$; since $L^\lambda(z) = \sum_n L^\lambda_n z^{-n-2}$ is a Virasoro field, it follows that ν^λ is a conformal vector (see also (5.1.10) below). Using also (3.6.16), we arrive at the following proposition.

PROPOSITION 5.1. *The vectors ν^λ ($\lambda \in \mathbb{C}$) are conformal vectors of the vertex algebra F. The field $Y\left(\nu^\lambda, z\right)$ is a Virasoro field with central charge*

$$c_\lambda = -12\lambda^2 + 12\lambda - 2.$$

The field $\psi^+(z)$ (resp. $\psi^-(z)$) is a primary field with respect to $Y\left(\nu^\lambda, z\right)$ of conformal weight λ (resp. $1 - \lambda$).

We turn now to bosonization (see Section 3.6). Let

$$\alpha(z) =: \psi^+(z)\psi^-(z) : .$$

This is an even field of conformal weight 1 with respect to any $L^\lambda(z)$. Due to (3.6.10) and (3.6.12) we have the following OPE:

$$(5.1.3) \qquad \alpha(z)\psi^\pm(w) \quad \sim \quad \pm\frac{\psi^\pm(w)}{z - w},$$

$$(5.1.4) \qquad \alpha(z)\alpha(w) \quad \sim \quad \frac{1}{(z - w)^2}.$$

Formula (5.1.4) shows that $\alpha(z)$ is a free bosonic field with affine central charge 1.

Furthermore in our case the second sum in (3.6.13) vanishes (due to Remark 3.3), hence (3.6.13) gives:

$$(5.1.5) \qquad : \alpha(z)\alpha(z) :=: \partial\psi^+(z)\psi^-(z) : + : \partial\psi^-(z)\psi^+(z) : .$$

It follows that

$$(5.1.6) \qquad Y\left(\nu^\lambda, z\right) \equiv L^\lambda(z) = \frac{1}{2} : \alpha(z)^2 : + \left(\frac{1}{2} - \lambda\right)\partial\alpha(z).$$

As usual, we write $\alpha(z) = \sum \alpha_n z^{n-1}$. Then (5.1.3) and (5.1.4) mean the following:

$$(5.1.7) \qquad \begin{aligned} [\alpha_m, \alpha_n] &= m\delta_{m,-n}, \\ [\alpha_m, \psi^\pm_{(n)}] &= \pm\psi^\pm_{(m+n)}. \end{aligned}$$

Thus, the α_m form the oscillator algebra $\mathfrak{s}$, and α_0, called the *charge operator*, is diagonalizable on F. The eigenvalues of α_0 are called *charges*. Explicitly, the elements

(5.1.8)

$$\psi^-_{(-j_t)} \cdots \psi^-_{(-j_1)} \psi^+_{(-i_s)} \cdots \psi^+_{(-i_1)} |0\rangle \quad (0 < i_1 < i_2 < \cdots, \quad 0 < j_1 < j_2 < \cdots)$$

are eigenvectors of α_0 of charge $s - t$ which form a basis of F.

Let $F = \oplus_{m \in \mathbb{Z}} F^{(m)}$ be the α_0-eigenspace decomposition, called the *charge decomposition*. Note that each $F^{(m)}$ is invariant with respect to $\mathfrak{s}$.

Furthermore, due to (5.1.6) we have, in particular:

(5.1.9)
$$L_0^\lambda = \frac{1}{2}\alpha_0^2 + \left(\lambda - \frac{1}{2}\right)\alpha_0 + \sum_{j=1}^\infty \alpha_{-j}\alpha_j.$$

Due to (5.1.2) we have:

(5.1.10)
$$[L_0^\lambda, \psi^+_{(m)}] = (-m - 1 + \lambda)\psi^+_{(m)},$$
$$[L_0^\lambda, \psi^-_{(m)}] = (-m - \lambda)\psi^-_{(m)}.$$

Hence L_0^λ, called the *energy operator*, is diagonalizable in the basis (5.1.8) of F, the eigenvalue, called the *energy*, of the element (5.1.8) being equal

(5.1.11)
$$i_1 + \cdots + i_s + j_1 + \cdots + j_t + \lambda(s - t) - s.$$

Note that the energy of all states is non-negative provided that $\lambda \in [0, 1]$.

Introduce the following element of $F^{(m)}$ called the m-th charged vacuum:

$$|m\rangle = \psi^+_{(-m)} \cdots \psi^+_{(-2)} \psi^+_{(-1)} |0\rangle \quad \text{if } m \geq 0,$$
$$|m\rangle = \psi^-_{(m)} \cdots \psi^-_{(-2)} \psi^-_{(-1)} |0\rangle \quad \text{if } m \leq 0.$$

It is easy now to prove the following important theorem.

THEOREM 5.1. *The representation of the oscillator algebra $\mathfrak{s}$ in each space $F^{(m)}$ is irreducible.*

PROOF. By Theorem 3.5b, it suffices to show that if $v \in F^{(m)}$ is a vector such that $\alpha_j v = 0$ for all $j > 0$, then $v \in \mathbb{C}|m\rangle$. It follows from (5.1.9) that v has the same energy as $|m\rangle$. But by (5.1.11) the vector $|m\rangle$ has the strictly lowest energy among the vectors (5.1.8) of charge m, if we take $\lambda \in (0, 1)$. $\square$

Here is a nice application of Theorem 5.1. Let us compute the "character"

$$\mathrm{ch}F = \mathrm{tr}_F q^{L_0^\lambda} z^{\alpha_0}$$

in two different ways. Just looking at the basis (5.1.8) we get

$$(5.1.12) \qquad \mathrm{ch}F = \prod_{j=1}^{\infty} \left(1 + zq^{\lambda+j-1}\right)\left(1 + z^{-1}q^{-\lambda+j}\right).$$

On the other hand, elements

$$\alpha_{-j_s}\cdots\alpha_{-j_1}|m\rangle \quad (0 < j_1 \le j_2 \le \cdots)$$

form a basis of L_0^λ-eigenvectors of $F^{(m)}$ with eigenvalues $m\lambda+\frac{1}{2}m(m-1)+j_1+\cdots j_s$. Hence

$$(5.1.13) \qquad \mathrm{ch}F = \sum_{m\in\mathbb{Z}} z^m q^{m\lambda+\frac{1}{2}m(m-1)} \Bigg/ \prod_{j=1}^{\infty}\left(1 - q^j\right).$$

Comparing (5.1.12) and (5.1.13) we get

$$(5.1.14) \qquad \prod_{j=1}^{\infty}\left(1 - q^j\right)\left(1 + zq^{j+\lambda-1}\right)\left(1 + z^{-1}q^{j-\lambda}\right) = \sum_{m\in\mathbb{Z}} z^m q^{m\lambda+\frac{1}{2}m(m-1)}.$$

Replacing in this formula zq^λ by $-z$ we get the famous Jacobi triple product identity:

$$(5.1.15) \qquad \prod_{j=1}^{\infty}\left(1 - q^j\right)\left(1 - zq^{j-1}\right)\left(1 - z^{-1}q^j\right) = \sum_{m\in\mathbb{Z}}(-z)^m q^{m(m-1)/2}.$$

This identity is the "denominator identity" for the affine Kac-Moody algebra $sl(2)^\wedge$. For each affine Kac-Moody algebra $\widehat{\mathfrak{g}}$ there exists a similar denominator identity, and all these are called Macdonald identities. They arise naturally in representation theory of affine algebras [**K2**, Chapter 12].

Letting in (5.1.14) $\lambda = \frac{1}{3}$ and $z = -1$ and replacing q by q^3 we get the no less famous Euler identity:

$$(5.1.16) \qquad \prod_{j=1}^{\infty}(1 - q^j) = \sum_{m\in\mathbb{Z}}(-1)^m q^{m(3m+1)/2}$$

Letting in (5.1.14) $\lambda = \frac{1}{2}$ and $z = -1$, and replacing q by q^2 we get another famous identity due to Gauss:

$$(5.1.17) \qquad \prod_{j=1}^{\infty}\frac{1 - q^j}{1 + q^j} = \sum_{m\in\mathbb{Z}}(-q)^{m^2}.$$

REMARK 5.1. Formula (5.1.13) can be rewritten as follows:

$$q^{-c_\lambda/24}\mathrm{ch}F = \sum_{m\in\mathbb{Z}} z^m q^{\frac{1}{2}\left(m+\lambda-\frac{1}{2}\right)^2} \Big/ \eta,$$

where $\eta = q^{\frac{1}{24}} \prod_{j=1}^{\infty}\left(1-q^j\right)$ is the Dedekind η-function. If we substitute $z = $ root of 1, $q = e^{2\pi i\tau}$, the right-hand side becomes a modular function in τ on the upper half-plane. This is a very general phenomenon in representation theory of affine algebras [**K2**, Chapter 13] and, more generally, vertex algebras [**Z**].

5.2. Boson-fermion correspondence and KP hierarchy

In the previous section, starting with charged free fermions $\psi^{\pm}(z)$, we constructed a free boson $\alpha(z)$. We wish now to express the fields $\psi^{\pm}(z)$ via the field $\alpha(z)$. This is obviously impossible since $\alpha(z)$ preserves charge whereas $\psi^{\pm}(z)$ changes charge by ± 1. For that reason we introduce a new (invertible) operator u on F, which changes charge, as follows. Consider the automorphism of the algebra C_{char} defined by

$$\psi^{+}_{(n)} \mapsto \psi^{+}_{(n-1)}, \quad \psi^{-}_{(n)} \mapsto \psi^{-}_{(n+1)}.$$

It is clear that this automorphism maps the annihilator (in C_{char}) of the vector $|m\rangle$ to that of the vector $|m+1\rangle$, $m \in \mathbb{Z}$. Hence there exists a unique invertible operator u on F such that

$$(5.2.1) \qquad u\psi^{\pm}_{(n)}u^{-1} = \psi^{\pm}_{(n\mp 1)}, \quad u|m\rangle = |m+1\rangle.$$

Since for $n \neq 0$ we have:

$$(5.2.2) \qquad \alpha_n = \sum_{i\in\mathbb{Z}} \psi^{+}_{(i)}\psi^{-}_{(n-i-1)},$$

(5.2.1) implies

$$(5.2.3) \qquad u\alpha_n u^{-1} = \alpha_n \quad \text{if } n \neq 0.$$

Since $u : F^{(m)} \to F^{(m+1)}$, we obtain

$$(5.2.4) \qquad u\alpha_0 u^{-1} = \alpha_0 - 1.$$

Now, due to (4.4.5) the field corresponding to the vector $|\pm m\rangle$ $(m > 0)$ of the vertex algebra F under the state-field correspondence is:

$$(5.2.5) \qquad Y(|\pm m\rangle, z) =: \partial^{(m-1)}\psi^\pm(z)\cdots\partial\psi^\pm(z)\psi^\pm(z):,$$

in particular,

$$(5.2.6) \qquad \psi^\pm(z) = Y(|\pm 1\rangle, z).$$

On the other hand, by the general OPE formula (4.6.2a) we have:

$$\alpha(z)Y(|m\rangle, w) \sim \sum_{j\geq 0} \frac{Y(\alpha_j|m\rangle, w)}{(z-w)^{j+1}},$$

and since $\alpha_j|m\rangle = \delta_{0,j}m|m\rangle$, we obtain

$$(5.2.7a) \qquad \alpha(z)Y(|m\rangle, w) \sim \frac{mY(|m\rangle, w)}{z-w},$$

or, equivalently,

$$(5.2.7b) \qquad [\alpha_j, Y(|m\rangle, w)] = mw^j Y(|m\rangle, w).$$

We also have, using (5.2.5):

(5.2.8)
$$Y(|m\rangle, z) : F^{(k)} \to F^{(k+m)}[[z, z^{-1}]], \quad |k\rangle \mapsto z^{mk}|m+k\rangle + \text{higher energy states}.$$

Formula (5.2.7b) along with (5.2.3 and 4) show that the field

$$X_m(z) := u^{-m} e^{m\sum_{j<0} \frac{z^{-j}}{j}\alpha_j} Y(|m\rangle, z) e^{m\sum_{j>0} \frac{z^{-j}}{j}\alpha_j}$$

commutes with all operators α_i $(i \in \mathbb{Z})$. Furthermore, due to (5.2.8), we have

$$X_m(z) : F^{(k)} \to F^{(k)}[[z, z^{-1}]], \quad |k\rangle \mapsto z^{mk}|k\rangle + \text{higher energy states}.$$

By Theorem 5.1, we conclude that

$$X_m(z)|_{F^{(k)}} = z^{mk} I_{F^{(k)}},$$

hence $X_m(z) = z^{m\alpha_0}$. We thus obtained the following remarkable formula:

$$(5.2.9) \qquad Y(|m\rangle, z) = u^m z^{m\alpha_0} e^{-m\sum_{j<0}\frac{z^{-j}}{j}\alpha_j} e^{-m\sum_{j>0}\frac{z^{-j}}{j}\alpha_j}.$$

REMARK 5.2. The field (5.2.9) first appeared in the early days of string theory under the name "vertex operator" in the following form (see the review [**Man**]). Consider the (multivalued) Veneziano field

$$\varphi(z) = q - ip \log z + i \sum_{n \neq 0} \frac{\alpha_n}{n} z^{-n},$$

so that $\alpha(z) = i\partial\varphi(z)$. Here the α_n with $n \neq 0$ form the Heisenberg algebra $[\alpha_m, \alpha_n] = m\delta_{m,-n}$, $p = \alpha_0$ is the momenta operator, and $u = e^{iq}$ where q is the conjugate coordinate operator (i.e., $[q, p] = i$). Then $: e^{mi\varphi(z)} :$ is the (well-defined) vertex operator, where the sign $: :$ of normal ordering means that p and α_n with $n > 0$ (i.e., operators that annihilate the vacuum) are moved to the right.

For $m \in \mathbb{Z}$ let $B^{(m)} = \mathbb{C}[x_1, x_2, \ldots]$ denote the representation space for the oscillator algebra given by the usual formulas (cf. Example 3.5):

$$\alpha_j = \frac{\partial}{\partial x_j} \quad \text{and} \quad \alpha_{-j} = jx_j \quad \text{for} \quad j > 0, \ \alpha_0 = mI.$$

We identify the space $B := \bigoplus_{m \in \mathbb{Z}} B^{(m)}$ with the space $\mathbb{C}[x_1, x_2, \ldots; u, u^{-1}]$ via the obvious identification $B^{(m)} = \mathbb{C}[x_1, x_2, \ldots]u^m$. Then the operator (5.2.9) looks on B as follows:

$$(5.2.10) \qquad \Gamma_m(z) = u^m z^{m\alpha_0} e^{m \sum_{j=1}^{\infty} z^{-j} x_j} e^{-m \sum_{j=1}^{\infty} \frac{z^{-j}}{j} \frac{\partial}{\partial x_j}}.$$

Thus the space B becomes a vertex algebra with the vacuum vector 1 and the following state-field correspondence (where, as usual, $\alpha(z) = \sum_j \alpha_j z^{-j-1}$):

$$Y(x_{j_1} \cdots x_{j_n} u^m, z) =: \partial^{j_1 - 1}\alpha(z) \cdots \partial^{j_n - 1}\alpha(z)\Gamma_m(z) : /j_1! \cdots j_n!.$$

We can state now the basic result, called the boson-fermion correspondence, which goes back to Skyrme [**Sk**]. Its proof follows from the above discussion and the uniqueness of representations of the oscillator algebra (see Example 3.5).

THEOREM 5.2. *There exists a unique isomorphism of vertex algebras $\sigma : F \overset{\sim}{\to} B$ such that*

$$\sigma|m\rangle = u^m, \ m \in \mathbb{Z} \quad and \quad \sigma\left(: \psi^+(z)\psi^-(z) :\right)\sigma^{-1} = \alpha(z).$$

Under this isomorphism we have for $m \geq 1$:

$$\sigma : \partial^{(m-1)}\psi^\pm(z) \cdots \partial\psi^\pm(z)\psi^\pm(z) : \sigma^{-1} = \Gamma_{\pm m}(z),$$

in particular:

$$\sigma\psi^{\pm}(z)\sigma^{-1} = \Gamma_{\pm 1}(z).$$

In what follows it will be convenient to write the fields $\psi^{\pm}(z)$ in the form

$$\psi^{+}(z) = \sum_{j\in\mathbb{Z}} \psi_j^{+} z^{-j-1}, \quad \psi^{-}(z) = \sum_{j\in\mathbb{Z}} \psi_j^{-} z^{-j},$$

so that

$$\psi_m^{\pm}\psi_n^{\mp} + \psi_n^{\mp}\psi_m^{\pm} = \delta_{m,-n}, \quad \psi_m^{\pm}\psi_n^{\pm} = -\psi_n^{\pm}\psi_m^{\pm}.$$

Using these fields one constructs a representation r of the Lie algebra gl_{∞} of all matrices $(a_{ij})_{i,j\in\mathbb{Z}}$ with a finite number of non-zero entries a_{ij} as follows. Let $E_{mn} = (\delta_{im}\delta_{jn})_{i,j\in\mathbb{Z}}$, $m,n \in \mathbb{Z}$, be the usual basis of gl_{∞}. We let

$$r(E_{mn}) = \psi_{-m}^{+}\psi_n^{-},$$

in other words:

$$(5.2.11) \qquad \sum_{i,j\in\mathbb{Z}} r(E_{ij})z^{i-1}w^{-j} = \psi^{+}(z)\psi^{-}(w) \equiv E(z,w).$$

We have

$$[E(z,w), \psi^{+}(u)] = \psi^{+}(z)\delta(w-u), \quad [E(z,w), \psi^{-}(u)] = -\psi^{-}(w)\delta(z-u),$$

in other words we have the following commutation relations:

$$(5.2.12) \qquad \left[r(E_{ij}), \psi_{-k}^{+}\right] = \delta_{jk}\psi_{-i}^{+}, \quad \left[r(E_{ij}), \psi_k^{-}\right] = -\delta_{ik}\psi_j^{-}.$$

It follows that the adjoint representation on C_{char} induces the defining representation of gl_{∞} in the space $\bigoplus_j \mathbb{C}\psi_{-j}^{+}$ (resp. its dual in the space $\bigoplus_j \mathbb{C}\psi_j^{-}$); in particular r is indeed a representation of gl_{∞}.

It is straightforward to see that the restriction of the representation r to each $F^{(m)}$ is irreducible and that this is the m-th fundamental representation of gl_{∞} with highest weight vector $|m\rangle$:

$$r(E_{ij})|m\rangle = 0 \text{ if } i < j; \quad r(E_{ii})|m\rangle = |m\rangle \text{ if } i \leq m, \text{ and } = 0 \text{ if } i > m.$$

Recall that we have (cf. (5.2.6) and (5.2.10))

$$(5.2.13) \qquad \sigma\psi^{\pm}(z)\sigma^{-1} = u^{\pm 1}z^{\pm\alpha_0}e^{\pm\sum_{j=1}^{\infty} z^{-j}x_j}e^{\mp\sum_{j=1}^{\infty}\frac{z^{-j}}{j}\frac{\partial}{\partial x_j}}.$$

Substituting this in (5.2.11) we obtain a formula for the representation of gl_∞ in $B^{(m)}$ (the bosonic picture):

$$\text{(5.2.14a)} \qquad \sum_{i,j\in\mathbb{Z}} \sigma r \sigma^{-1}(E_{ij}) z^{i-1} w^{-j} = \frac{(z/w)^m}{z-w}\Gamma(z,w),$$

where

$$\text{(5.2.14b)} \qquad \Gamma(z,w) = e^{\sum_{j=1}^{\infty}(z^j-w^j)x_j} e^{-\sum_{j=1}^{\infty}\frac{z^{-j}-w^{-j}}{j}\frac{\partial}{\partial x_j}}.$$

Formula (5.2.14a) should be understood as an equality of formal distributions in the domain $|z| > |w|$. Note also that in multiplying out of the vertex operators we have used

$$\text{(5.2.15)} \qquad e^{a\partial_x}e^{bx} = e^{ab}e^{bx}e^{a\partial_x}, \quad a,b\in\mathbb{C}.$$

One of the most remarkable applications of the boson-fermion correspondence is the theory of the KP hierarchy developed by the Kyoto school. We discuss this briefly, referring to [**DJKM**], [**JM**], [**KR**], or [**KL3**] for details.

The KP hierarchy in the fermionic picture is the following equation on $\tau \in F^{(0)}$:

$$\text{(5.2.16)} \qquad \sum_{j\in\mathbb{Z}} \psi_j^+\tau \otimes \psi_{-j}^-\tau = 0 \quad (\text{in } F\otimes F).$$

Introducing the operator

$$S = \sum_{j\in\mathbb{Z}} \psi_j^+ \otimes \psi_{-j}^-$$

on $F\otimes F$, we can rewrite (5.2.16) as

$$\text{(5.2.17)} \qquad S(\tau\otimes\tau) = 0.$$

Yet another way to rewrite (5.2.16) is

$$\text{(5.2.18)} \qquad \text{Res}_z \, \psi^+(z)\tau \otimes \psi^-(z)\tau = 0.$$

The two basic properties of the operator S are

$$\text{(5.2.19)} \qquad S(|0\rangle \otimes |0\rangle) = 0,$$

(i.e., the vacuum vector $\tau = |0\rangle$ is a solution of (5.2.16)) and

$$\text{(5.2.20)} \qquad [E(z,w)\otimes 1 + 1\otimes E(z,w), S] = 0,$$

(i.e., the representation of gl_∞ in $F \otimes F$ commutes with the operator S). Formula (5.2.19) is clear since either $\psi_j^- |0\rangle = 0$ or $\psi_{-j}^+ |0\rangle = 0$ for each j. Formula (5.2.20) is obtained by a simple calculation:

$$
\begin{aligned}
&\left[E(u,w) \otimes 1 + 1 \otimes E(u,w), \mathrm{Res}_z\, \psi^+(z) \otimes \psi^-(z) \right] \\
&= \ \mathrm{Res}_z\, \left[\psi^+(u)\psi^-(w), \psi^+(z) \right] \otimes \psi^-(z) + \mathrm{Res}_z\, \psi^+(z) \otimes \left[\psi^+(u)\psi^-(w), \psi^-(z) \right] \\
&= \ \mathrm{Res}_z\, \delta(w-z)\psi^+(u) \otimes \psi^+(z) - \mathrm{Res}_z\, \psi^+(z) \otimes \delta(u-z)\psi^-(w) \\
&= \ \psi^+(u) \otimes \psi^-(w) - \psi^+(u) \otimes \psi^-(w) = 0.
\end{aligned}
$$

The representation r of the Lie algebra gl_∞ exponentiates to a representation R of the group GL_∞ of all invertible matrices $(\delta_{ij} + a_{ij})_{i,j \in \mathbb{Z}}$ with a finite number of non-zero a_{ij}. Property (5.2.20) means that the operators $R(g) \otimes R(g)$ $(g \in GL_\infty)$ commute with the operator S on $F \otimes F$. Hence, applying $R(g) \otimes R(g)$ to both sides of equation (5.2.19), we obtain that all elements $\tau = R(g) \cdot |0\rangle$ $(g \in GL_\infty)$ are solutions of the KP hierarchy. (One can show that, conversely, if a non-zero element τ of F is a solution of the KP hierarchy then τ lies on the orbit $R(GL_\infty)|0\rangle$ [**KR**].)

Let us go now to the bosonic picture. We identify $F^{(0)}$ with $B^{(0)} = \mathbb{C}[x_1, x_2, \dots]$ using σ, and $F^{(0)} \otimes F^{(0)}$ with $B^{(0)} \otimes B^{(0)} = \mathbb{C}[x_1', x_2', \dots \,; x_1'', x_2'', \dots]$. Substituting in (5.2.18) the right-hand side of (5.2.13) we obtain the "bilinear form" of the KP hierarchy in the bosonic picture:

$$
(5.2.21) \qquad \mathrm{Res}_z\, \left(e^{z \cdot x'} e^{-\sum_{j=1}^{\infty} \frac{z^{-j}}{j} \frac{\partial}{\partial x_j'}} \tau(x') \right) \left(e^{-z \cdot x''} e^{\sum_{j=1}^{\infty} \frac{z^{-j}}{j} \frac{\partial}{\partial x_j''}} \tau(x'') \right) = 0,
$$

where $x = (x_1, x_2, \dots)$ and $z \cdot x$ stands for $\sum_{j=1}^{\infty} z^j x_j$.

There are two ways to proceed from (5.2.21). The first way is to introduce new variables by letting $x' = x - y$, $x'' = x + y$, which leads to the KP hierarchy of Hirota bilinear equations on the τ-function. We refer to the literature quoted above for details. The second way is to introduce the wave functions

$$
w^\pm(x,z) = \frac{e^{\pm z \cdot x} e^{\mp \sum_{j=1}^{\infty} \frac{z^{-j}}{j} \frac{\partial}{\partial x_j}} \tau(x)}{\tau(x)},
$$

so that equation (5.2.18) becomes

$$
(5.2.22) \qquad\qquad\qquad \mathrm{Res}_z\, w^+(x', z) w^-(x'', z) = 0.
$$

The wave functions $w^\pm(x,z)$ have the following form: $w^\pm(x,z) = (1 + \sum_{j=1}^\infty w_j^\pm(x)z^{-j})e^{\pm z\cdot x}$. Introduce the wave operators $P^\pm = 1 + w_1^\pm(x)\partial^{-1} + w_2^\pm\partial^{-2} + \cdots$, so that $w^\pm(x,z) = P^\pm e^{\pm z\cdot x}$ and let $L = P^+\partial(P^+)^{-1}$, where $\partial = \partial_{x_1}$. Then $L = \partial + u_1(x)\partial^{-1} + u_2(x)\partial^{-2} + \cdots$, where

$$u_1 = \frac{\partial^2}{\partial x_1^2}\log\tau(x)\,, \text{ etc.}$$

One can show that (5.2.22) implies the following hierarchy of evolution equations of L:

$$(5.2.23) \qquad \frac{\partial L}{\partial x_n} = \left[(L^n)_+\,, L\right], \quad n = 1, 2,\ldots,$$

where the subscript $+$ signifies the differential part of a pseudo-differential operator. Equations (5.2.23) imply the following zero curvature equations:

$$(5.2.24) \qquad \left[\frac{\partial}{\partial x_m} - (L^m)_+\,, \frac{\partial}{\partial x_n} - (L^n)_+\right] = 0, \quad m, n = 1, 2,\ldots\,.$$

Equation (5.2.24) for $m = 2$, $n = 3$ produces the classical KP equation on the function $u = 2u_1$, where $x_1 = x$, $x_2 = y$, $x_3 = t$:

$$(5.2.25) \qquad \frac{3}{4}u_{yy} = \left(u_t - \frac{3}{2}uu_x - \frac{1}{4}u_{xxx}\right)_x.$$

One can show that if $\tau(x)$ is a solution of (5.2.21) then $(1 + \alpha\Gamma(a,b))\tau(x)$ is a solution as well for any $\alpha, a, b \in \mathbb{C}$. Applying this procedure N times starting with $\tau(x) = 1$ we obtain the τ-function of the so called N-soliton solution. For example, $u(x,y,t) = 2\frac{\partial^2}{\partial x^2}\log\big(1+\Gamma(a,b)\big)\tau(x,y,t))$ is a 1-soliton solution of (5.2.25); explicitly:

$$u(x,y,t) = \frac{1}{2}(a-b)^2\left(\cosh\frac{1}{2}((a-b)x + (a^2-b^2)y + (a^3-b^3)t + \text{const})\right)^{-2}.$$

5.3. $\widehat{gl}_\infty$ and $W_{1+\infty}$

Denote by $\widetilde{gl}_\infty$ the Lie algebra of all matrices $(a_{ij})_{i,j\in\mathbb{Z}}$ such that $a_{ij} = 0$ for $|i - j| \gg 0$. It is important to consider this Lie algebra, which is larger than gl_∞, because, as we shall see, many important Lie algebras can be embedded in $\widetilde{gl}_\infty$ but not in gl_∞.

Unfortunately the representation r can not be extended from gl_∞ to $\widetilde{gl}_\infty$, since for example,

$$r\left(\mathrm{diag}(\lambda_i)_{i\in\mathbb{Z}}\right)|0\rangle = \sum_{i\le 0}\lambda_i|0\rangle.$$

In order to remove this "anomaly," introduce the following projective representation $\widehat{r}$ of the Lie algebra gl_∞ (cf. (5.2.11)):

$$(5.3.1) \qquad \sum_{i,j\in\mathbb{Z}}\widehat{r}(E_{ij})\,z^{i-1}w^{-j} =: \psi^+(z)\psi^-(w) :$$

It is clear that $\widehat{r}$ extends to a projective representation of the Lie algebra $\widetilde{gl}_\infty$. Recall that in the domain $|z| > |w|$ we have

$$\psi^+(z)\psi^-(w) = \frac{1}{z-w} + :\psi^+(z)\psi^-(w): .$$

Comparing this with (5.2.11) and (5.3.1) we obtain:

$$(5.3.2) \qquad \begin{aligned} \widehat{r}(E_{ij}) &= r(E_{ij}) \qquad \text{if } i\ne j \text{ or } i=j>0, \\ \widehat{r}(E_{ii}) &= r(E_{ii}) - I \quad \text{if } i\le 0. \end{aligned}$$

It follows that

$$[\widehat{r}(A),\widehat{r}(B)] = \widehat{r}([A,B]) + \alpha(A,B)I, \quad A,B\in\widetilde{gl}_\infty,$$

where $\alpha(A,B) : \widetilde{gl}_\infty \times \widetilde{gl}_\infty \to \mathbb{C}$ is a bilinear function given by

$$(5.3.3) \qquad \alpha(A,B) = \mathrm{tr}\left([J,A]B\right), \qquad \text{where } J = \sum_{i\le 0} E_{ii}.$$

Since $\widetilde{gl}_\infty$ is a Lie algebra, it follows that $\alpha(A,B)$ is skewsymmetric and satisfies the identity

$$\alpha\left([A,B],C\right) + \alpha\left([B,C],A\right) + \alpha\left([C,A],B\right) = 0,$$

i.e., α is a 2-cocycle on $\widetilde{gl}_\infty$.

Let $\widehat{gl}_\infty = \widetilde{gl}_\infty + \mathbb{C}K$ be the central extension of $\widetilde{gl}_\infty$ defined by this cocycle, that is K is a central element and the Lie algebra bracket on $\widehat{gl}_\infty$ for any two elements A, B from the subspace $\widetilde{gl}_\infty$ is given by

$$[A,B] = AB - BA + \alpha(A,B)K.$$

Thus, letting $\widehat{r}(K) = I$, we obtain a linear representation $\widehat{r}$ in the vector space F of the Lie algebra $\widehat{gl}_\infty$, defined on the subspace $\widetilde{gl}_\infty$ by (5.3.2).

Consider now the Lie algebra $\mathcal{D} = \text{Diff } \mathbb{C}^\times$ of regular differential operators on $\mathbb{C}^\times = \mathbb{C} \setminus \{0\}$ (cf. Section 2.10). Recall that these operators are of the form

$$\sum_{j=0}^{N} a_j(t)\partial_t^j, \qquad \text{where } a_j(t) \in \mathbb{C}[t, t^{-1}].$$

They act on $\mathbb{C}[t, t^{-1}]$ in the usual way. Operators

$$J_n^k = t^{k+n}(-\partial_t)^k \ (k \in \mathbb{Z}_+, n \in \mathbb{Z})$$

form a basis of $\mathcal{D}$.

Choosing the basis $e_j = t^{-j}$ of $\mathbb{C}[t, t^{-1}]$, we obtain

$$\frac{1}{k!} J_n^k e_j = (-1)^k \binom{-j}{k} e_{j-n}.$$

We thus get the following embedding φ of $\mathcal{D}$ in $\widetilde{gl}_\infty$:

$$(5.3.4) \qquad \varphi\left(\frac{1}{k!} J_n^k\right) = (-1)^k \sum_{j \in \mathbb{Z}} \binom{-j}{k} E_{j-n,j}.$$

Let $\widehat{\mathcal{D}} = \mathcal{D} + \mathbb{C}C$ denote the central extension of the Lie algebra $\mathcal{D}$ defined by the cocycle α restricted to $\varphi(\mathcal{D})$, and extend φ to a homomorphism $\widehat{\varphi} : \widehat{\mathcal{D}} \to \widehat{gl}$ by letting $\widehat{\varphi}(C) = K$.

Introduce the following formal distributions with values in $\widehat{\mathcal{D}}$ ($k \in \mathbb{Z}_+$):

$$J^k(z) = \sum_{n \in \mathbb{Z}} J_n^k z^{-k-n-1}.$$

Then (5.3.4) becomes:

$$(5.3.5) \qquad \frac{1}{k!} \widehat{\varphi}(J^k(z)) = (-1)^k \sum_{i,j \in \mathbb{Z}} \binom{-j}{k} E_{ij} z^{i-j-k-1}.$$

THEOREM 5.3. *(a)* $\widehat{r}\left(\widehat{\varphi}\left(J^k(z)\right)\right) = (-1)^k : \psi^+(z)\partial^k \psi^-(z):$.

(b) The formal distributions $J^k(z)$ are mutually local.

(c) The restriction of the cocycle α to $\mathcal{D}$ via the embedding $\widehat{\varphi}$ is given by the following formula:

$$\alpha\left(f(t)\partial_t^r, g(t)\partial_t^s\right) = \frac{r! s!}{(r+s+1)!} \text{Res}_t \left(\partial_t^{s+1} f(t)\right) \left(\partial_t^r g(t)\right).$$

PROOF. Differentiating both sides of (5.3.1) k times by w and letting $w = z$, we see that (a) follows from (5.3.5); (b) follows from (a). Note that, by Wick's formula, the constant term in the OPE $: \psi^+(z)\partial^r\psi^-(z) : : \psi^+(w)\partial^s\psi^-(w) :$ is equal to $\frac{(-1)^r r! s!}{(z-w)^{r+s+2}}$. Hence, using (2.6.2a) and (a), we obtain

$$(5.3.6) \qquad \alpha\left(J_m^r, J_n^s,\right) = (-1)^s r! s! \binom{m+r}{r+s+1} \delta_{m,-n},$$

proving (c). $\qquad\qquad\qquad\qquad\qquad\qquad\qquad\qquad\qquad\qquad\qquad\qquad\qquad\qquad\square$

Recall that we have proved the locality of formal distributions $J^k(z)$ already in Section 2.10 by a direct calculation.

REMARK 5.3a. The restriction of the cocycle α to the subalgebra gl_∞ of the Lie algebra $\widetilde{gl}_\infty$ produces a trivial cocycle:

$$\alpha(A, B) = \mathrm{tr}\, J[A, B].$$

On the other hand,

$$\alpha(t^m, t^n) = m\delta_{m,-n},$$

hence restricted already to the (commutative) subalgebra $\varphi\left(\sum_n \mathbb{C}t^n\right)$, the cocycle α is nontrivial. Note also that when restricted to vector fields, α reproduces a multiple of the Virasoro cocycle:

$$\alpha\left(t^{m+1}\partial_t, t^{n+1}\partial_t\right) = -\delta_{m,-n}\frac{m^3 - m}{6}.$$

It is easy to see that the derivation $T = -ad\partial_t$ of the Lie algebra $\mathcal{D}$ lifts to the central extension $\widehat{\mathcal{D}}$ by letting $T(C) = 0$. In fact, this is equivalent to the relation

$$\alpha\left([\partial_t, A], B\right) = -\alpha\left(A, [\partial_t, B]\right), \quad A, B \in \mathcal{D},$$

which is immediate to check. It is also immediate to check that

$$T(J^k(z)) = \partial J^k(z), \quad k \in \mathbb{Z}_+.$$

Hence we may apply the construction of vertex algebras associated to regular formal distribution Lie algebras developed in Section 4.7.

We have:

$$\widehat{\mathcal{D}}_{--} = \mathcal{P} + \mathbb{C}C, \qquad T\widehat{\mathcal{D}}_{--} = \mathcal{P},$$

where $\mathcal{P}$, the annihilation subalgebra of $\widehat{\mathcal{D}}$, consists of all differential operators regular on the whole complex plane $\mathbb{C}$ (i.e., the coefficients of these operators are in $\mathbb{C}[t]$). Note that $\mathcal{P}$ is a subalgebra of $\widehat{\mathcal{D}}$ since the restriction of the cocycle α to $\mathcal{P}$ is zero. Hence all 1-dimensional $\widehat{\mathcal{D}}_{--}$ – modules λ are of the form:

$$\lambda_c(\mathcal{P}) = 0, \qquad \lambda_c(C) = c \in \mathbb{C}.$$

We thus obtain the universal vertex algebras (see Theorem 4.7):

$$\widetilde{V}^c(\widehat{\mathcal{D}}) = \mathrm{Ind}_{\mathcal{P}+\mathbb{C}C}^{\widehat{\mathcal{D}}} \lambda_c, \quad c \in \mathbb{C},$$

strongly generated by the fields $J^k(z)$, $k \in \mathbb{Z}_+$.

Finally, it is straightforward to check that for each $\lambda \in \mathbb{C}$, the vector

$$(5.3.7) \qquad \nu^\lambda = \left(J^1_{-2} + (1-\lambda) J^0_{-2} \right) |0\rangle$$

is a conformal vector of the vertex algebra $\widetilde{V}^c(\widehat{\mathcal{D}})$. Note that the corresponding Virasoro field is

$$Y\left(\nu^\lambda, z\right) = J^1(z) + (1-\lambda)\, \partial J^0(z),$$

and that the central charge equals

$$(5.3.8) \qquad -\left(12\lambda^2 - 12\lambda + 2\right) c.$$

In particular, $\widetilde{V}^c(\widehat{\mathcal{D}})$ is a graded vertex algebra (with the Hamiltonian J^1_0), hence it has a unique simple quotient $V^c(\widehat{\mathcal{D}})$. The generally accepted notation for the simple vertex algebra $V^c(\widehat{\mathcal{D}})$ is $W_{1+\infty,c}$. More on these vertex algebras and references to their applications may be found in [**FKRW**] and [**KRad**].

REMARK 5.3b. Due to Theorem 5.1, the fields $: \psi^+(z)\partial^k\psi^-(z) :$, $k \in \mathbb{Z}_+$, generate the subalgebra $F^{(0)}$ of the vertex algebra F (it suffices to take $k = 0$). It follows from Theorem 5.3a that the vertex algebra $F^{(0)}$ is isomorphic to the vertex algebra $W_{1+\infty,1}$. The conformal vectors (5.3.7) and (5.1.1) correspond under this isomorphism.

REMARK 5.3c. Consider the Lie algebra of $N \times N$ matrix-valued differential operators on $\mathbb{C}^\times : \mathcal{D}^N = \mathrm{gl}_N(\mathrm{Diff}\,\mathbb{C}^\times)$. The cocycle α given by (5.3.6) extends to the Lie algebra $\mathcal{D}^N$ by the formula

$$\alpha\left(J^r_m A, J^s_n B\right) = \alpha\left(J^r_m, J^s_n\right) \mathrm{tr}\, AB, \quad A, B \in \mathrm{gl}_N(\mathbb{C}).$$

The corresponding central extension $\widehat{\mathcal{D}}^N$ is a regular formal distribution Lie algebra (cf. Section 2.10). Its annihilation subalgebra $\mathcal{P}^N$ consists of all regular (matrix valued) differential operators on $\mathbb{C}$. As above, we have, for each $c \in \mathbb{C}$, the universal vertex algebra $\tilde{V}^c(\widehat{\mathcal{D}}^N)$ and its unique simple quotient $V^c(\widehat{\mathcal{D}}^N)$.

5.4. Lattice vertex algebras

The vertex algebra $B^k(\mathbb{R}^\ell)$ of ℓ free bosons may be viewed as a quantization of the space of maps from the circle S^1 to $\mathbb{R}^\ell$. In this section, we shall construct a vertex algebra V_Q associated to an integral lattice Q of rank ℓ which may be viewed as a quantization of the space of maps from S^1 to the torus $\mathbb{R}^\ell/Q$. This is called a *lattice vertex algebra*.

Let Q be a free abelian group of rank ℓ. Recall that the group algebra $\mathbb{C}[Q]$ is an algebra with basis $e^\alpha (\alpha \in Q)$ and multiplication

$$e^\alpha e^\beta = e^{\alpha+\beta}, \quad e^0 = 1 \qquad (\alpha, \beta \in Q).$$

Let Q be given a structure of an integral lattice, meaning that Q is equipped with a $\mathbb{Z}$-valued symmetric bilinear form $(.|.)$. Let $\mathfrak{h} = \mathbb{C} \otimes_{\mathbb{Z}} Q$ be the complexification of Q, and extend the bilinear form $(.|.)$ from Q to $\mathfrak{h}$ by bilinearity. Let

$$\widehat{\mathfrak{h}} = \mathfrak{h}\left[t, t^{-1}\right] + \mathbb{C}K$$

be the affinization of $\mathfrak{h}$ viewed as a commutative Lie algebra (see Section 3.5). Let S be the symmetric algebra over the space $\mathfrak{h}^{<0} = \sum_{j<0} \mathfrak{h} \otimes t^j$. We shall write ht^j in place of $h \otimes t^j$ for short.

We define the space of states of the vertex algebra that we shall associate to the lattice Q as

$$V_Q = S \otimes \mathbb{C}[Q]$$

with the parity

$$p(s \otimes e^\alpha) = p(\alpha) \in \mathbb{Z}/2\mathbb{Z},$$

where $p : Q \to \mathbb{Z}/2\mathbb{Z}$ is a homomorphism to be determined, and the vacuum vector

$$|0\rangle = 1 \otimes 1.$$

Recall (see Section 3.5) that we have a representation, which will be denoted by π_1, of the Lie algebra $\widehat{\mathfrak{h}}$ in the space S defined by letting $\pi_1(K) = I$, $\pi_1(ht^n)$

be the operator of multiplication by ht^n if $n < 0$, $\pi(ht^n)$ be the derivation of the algebra S defined by

$$(ht^n)(at^{-s}) = n\delta_{n,s}(h|a)$$

if $n > 0$ and $\pi_1(h) = 0$ $(h, a \in \mathfrak{h}, s > 0)$.

Recall that the fields $\sum_{n \in \mathbb{Z}} \pi_1(ht^n)z^{-n-1}$ generate a vertex algebra structure on the space S. In order to extend this structure to V_Q, we define a representation π_2 of $\widehat{\mathfrak{h}}$ on the space $\mathbb{C}[Q]$ by letting

$$\pi_2(K) = 0, \quad \pi_2(ht^n)e^\alpha = \delta_{n,0}\,(\alpha|h)\,e^\alpha \quad (h \in \mathfrak{h},\, \alpha \in Q,\, n \in \mathbb{Z})\,,$$

and extend π_1 to a representation π of $\widehat{\mathfrak{h}}$ on V_Q by $\pi = \pi_1 \otimes 1 + 1 \otimes \pi_2$. Let $h_n = \pi(ht^n)$ $(h \in \mathfrak{h}, n \in \mathbb{Z})$ and consider the following $\mathrm{End}\,V_Q$-valued fields: $h(z) = \sum_{n \in \mathbb{Z}} h_n z^{-n-1}$. Then we have

$$(5.4.1a) \qquad [h_m, h'_n] = m\delta_{m,-n}\,(h|h')\,, \quad h, h' \in \mathfrak{h}, \quad m, n \in \mathbb{Z}\,,$$

which is equivalent to the OPE

$$(5.4.1b) \qquad h(z)h'(w) \sim \frac{(h|h')}{(z-w)^2}\,.$$

Denoting by e^α the operator on V_Q of multiplication by $1 \otimes e^\alpha (\alpha \in Q)$, we have

$$(5.4.2) \qquad [h_n, e^\alpha] = \delta_{n,0}\,(\alpha|h)\,e^\alpha, \quad n \in \mathbb{Z}, h \in \mathfrak{h}\,.$$

In order to construct the state-field correspondence, we need to find the fields $\Gamma_\alpha(z) := Y(1 \otimes e^\alpha, z)$ for each $\alpha \in Q$. Since $h_n e^\alpha|0\rangle = \delta_{n,0}(h|\alpha)e^\alpha|0\rangle$ for $h \in \mathfrak{h}$ and $n \in \mathbb{Z}_+$, we see from the general OPE formula (4.6.2a) that we must have

$$(5.4.3a) \qquad h(z)\Gamma_\alpha(w) \sim \frac{(\alpha|h)\Gamma_\alpha(w)}{z-w} \quad \text{for } h \in \mathfrak{h},\, \alpha \in Q\,,$$

which is equivalent to

$$(5.4.3b) \qquad [h_n, \Gamma_\alpha(w)] = (\alpha|h)w^n\Gamma_\alpha(w) \quad \text{for } h \in \mathfrak{h},\, n \in \mathbb{Z},\, \alpha \in Q\,.$$

Using the same argument as in the proof of formula (5.2.9), we derive from (5.4.3b) that

$$\Gamma_\alpha(z) = e^\alpha e^{-\sum_{j<0} \frac{z^{-j}}{j}\alpha_j} e^{-\sum_{j>0} \frac{z^{-j}}{j}\alpha_j} a_\alpha(z)\,,$$

where $a_\alpha(z)$ is a field such that

$$(5.4.4) \qquad [h_n, a_\alpha(z)] = 0 \quad \text{for all } h \in \mathfrak{h},\ n \in \mathbb{Z}.$$

Furthermore, we want the fields $\Gamma_\alpha(z)$, $\alpha \in Q$, to be pairwise local. In the same way as in the derivation of (5.2.14b) we obtain in the domain $|z| > |w|$:

$$(5.4.5a) \qquad \Gamma_\alpha(z)\Gamma_\beta(w) = e^\alpha a_\alpha(z) e^\beta a_\beta(w) \left(1 - \frac{w}{z}\right)^{(\alpha|\beta)} c_{\alpha,\beta}(z,w),$$

where

$$(5.4.5b) \qquad c_{\alpha,\beta}(z,w) = e^{-\sum_{j<0}\left(\frac{z^{-j}}{j}\alpha_j + \frac{w^{-j}}{j}\beta_j\right)} e^{-\sum_{j>0}\left(\frac{z^{-j}}{j}\alpha_j + \frac{w^{-j}}{j}\beta_j\right)}.$$

As before, we have used the formula

$$e^{a\alpha_m} e^{b\beta_n} = e^{ab[\alpha_m,\beta_n]} e^{b\beta_n} e^{a\alpha_m}, \quad a,b \in \mathbb{C},\ \alpha,\beta \in \mathfrak{h}.$$

Using the equivalent definition of locality given by Theorem 2.3(vii), we conclude that the fields $\Gamma_\alpha(z)$ and $\Gamma_\beta(w)$ are mutually local iff the following equality holds for all z,w:

$$(5.4.6) \quad e^\alpha a_\alpha(z) e^\beta a_\beta(w) z^{-(\alpha|\beta)} = (-1)^{p(\alpha)p(\beta)+(\alpha|\beta)} e^\beta a_\beta(w) e^\alpha a_\alpha(z) w^{-(\alpha|\beta)}.$$

Furthermore, we have

$$\Gamma_\alpha(z)|0\rangle = e^\alpha a_\alpha(z) e^{\sum_{j>0}\frac{z^j}{j}\alpha_{-j}}|0\rangle = e^\alpha a_\alpha(z)\left(1 + z\alpha_{-1} + \cdots\right)|0\rangle.$$

Hence by the vacuum axiom we must have

$$(5.4.7) \qquad a_\alpha(z)|0\rangle\,|_{z=0} = |0\rangle, \quad a_0(z) = 1,$$

and also we must have (see (1.3.3))

$$(5.4.8) \qquad T(1 \otimes e^\alpha) = \left(\alpha t^{-1}\right) \otimes e^\alpha.$$

Since we want (4.4.6) for $n = 0$ and (4.4.7) to hold, formula (5.4.8) forces

$$(5.4.9) \qquad \partial\Gamma_\alpha(z) =: \alpha(z)\Gamma_\alpha(z):\,.$$

The latter equation is equivalent to $\partial a_\alpha(z) = \alpha_0 z^{-1} a_\alpha(z)$, hence we must have

$$a_\alpha(z) = c_\alpha z^{\alpha_0},$$

where c_α is an operator independent of z such that due to (5.4.4) and (5.4.7):

$$(5.4.10) \qquad c_0 = 1, \quad c_\alpha|0\rangle = |0\rangle, \quad [h_n, c_\alpha] = 0 \quad (h \in \mathfrak{h},\ n \in \mathbb{Z}).$$

Using (5.4.2), we see that the locality condition (5.4.6) is equivalent to

$$(5.4.11) \qquad e^{\alpha} c_{\alpha} e^{\beta} c_{\beta} = (-1)^{p(\alpha)p(\beta)+(\alpha|\beta)} e^{\beta} c_{\beta} e^{\alpha} c_{\alpha} \,.$$

It is also clear that all $\Gamma_{\alpha}(z)$ are indeed fields. We thus arrive at the following proposition (by making use of Theorem 4.5):

PROPOSITION 5.4. *Any vertex algebra with the space of states V_Q and the vacuum vector $|0\rangle = 1 \otimes 1$ with the property*

$$(5.4.12) \qquad Y\left(\left(ht^{-1}\right) \otimes 1, z\right) = h(z) \quad \text{for all } h \in \mathfrak{h},$$

is generated by the fields $h(z)$ $(h \in \mathfrak{h})$ and the fields

$$Y\left(1 \otimes e^{\alpha}, z\right) = e^{\alpha} z^{\alpha_0} e^{-\sum_{j<0} \frac{z^{-j}}{j} \alpha_j} e^{-\sum_{j>0} \frac{z^{-j}}{j} \alpha_j} c_{\alpha} \quad (\alpha \in Q),$$

where the c_{α} are operators on V_Q satisfying conditions (5.4.10) and (5.4.11). For any solution of equations (5.4.10) and (5.4.11), there exists a unique such vertex algebra.

The most important solutions to the equations (5.4.10) and (5.4.11) are of the following form (these are all solutions if $(\cdot \mid \cdot)$ is non-degenerate):

$$(5.4.13) \qquad c_{\alpha}(s \otimes e^{\beta}) = \epsilon(\alpha, \beta) s \otimes e^{\beta} \quad (s \in S,\ \beta \in Q),$$

where $\epsilon(\alpha, \beta) \in \mathbb{C}$. Equations (5.4.10) and (5.4.11) are then equivalent to

$$(5.4.14a) \qquad \epsilon(\alpha, 0) = \epsilon(0, \alpha) = 1 \qquad (\alpha \in Q),$$

$$(5.4.14b) \qquad \epsilon(\alpha, \beta) = (-1)^{p(\alpha)p(\beta)+(\alpha|\beta)} \epsilon(\beta, \alpha) \qquad (\alpha, \beta \in Q),$$

$$(5.4.14c) \qquad \epsilon(\beta, \gamma)\epsilon(\beta+\gamma, \alpha) = \epsilon(\beta, \alpha+\gamma)\epsilon(\gamma, \alpha) \qquad (\alpha, \beta, \gamma \in Q).$$

Indeed, (5.4.14b) follows from (5.4.11) applied to the vacuum vector and using (5.4.10). Since the function

$$B(\alpha, \beta) = (-1)^{p(\alpha)p(\beta)+(\alpha|\beta)} \quad (\alpha, \beta \in Q)$$

is bimultiplicative, we see that the equation (5.4.11) for the c_{α} of the form (5.4.13) is equivalent to the equations (5.4.14b and c).

5.5. Simple lattice vertex algebras

In order to understand better the equations (5.4.14a–c), introduce the *twisted group algebra* $\mathbb{C}_\epsilon[Q]$. This is the algebra with a basis e^α $(\alpha \in Q)$ and the "twisted" multiplication:

$$e^\alpha e^\beta = \epsilon(\alpha, \beta)e^{\alpha+\beta} \quad (\alpha, \beta \in Q)\,.$$

Then equations (5.4.14a and c) simply mean that $\mathbb{C}_\epsilon[Q]$ is an associative algebra with the unit element $e^0 = 1$.

Note that e^α is an invertible element of the algebra $\mathbb{C}_\epsilon[Q]$ iff $\epsilon(\alpha, -\alpha) \neq 0$. Let $Q_\epsilon = \{\alpha \in Q | \epsilon(\alpha, -\alpha) \neq 0\}$ and let J_ϵ denote the linear span of e^α's such that $\epsilon(\alpha, -\alpha) = 0$. Then Q_ϵ is a sublattice of Q, J_ϵ is an ideal of the algebra $\mathbb{C}_\epsilon[Q]$ such that

$$(5.5.1) \qquad \mathbb{C}_\epsilon[Q_\epsilon] \simeq \mathbb{C}_\epsilon[Q]/J_\epsilon\,.$$

Note also that

$$(5.5.2) \qquad \epsilon(\alpha, \beta)\epsilon(\beta, \alpha) \neq 0 \quad \text{for all } \beta \in Q \text{ if } \alpha \in Q_\epsilon\,.$$

Suppose now that $Q = Q_\epsilon$, or, equivalently, that $\epsilon : Q \times Q \to \mathbb{C}^\times$. Then equations (5.4.14a and c) mean that ϵ is a 2-cocycle of the group Q with values in the group $\mathbb{C}^\times$. Given a 2-cocycle $\epsilon : Q \times Q \to \mathbb{C}^\times$, one associates to ϵ a function $B_\epsilon : Q \times Q \to \mathbb{C}^\times$ defined by

$$(5.5.3) \qquad B_\epsilon(\alpha, \beta) = \epsilon(\alpha, \beta)\epsilon(\beta, \alpha)^{-1}\,.$$

It is clear that B_ϵ is skewsymmetric, i.e.,

$$B_\epsilon(\alpha, \beta) = B_\epsilon(\beta, \alpha)^{-1}\,.$$

Since (5.5.3) is equivalent to

$$e^\alpha e^\beta = B_\epsilon(\alpha, \beta)e^\beta e^\alpha\,,$$

multiplying both sides of this equality by e^γ on the right and using associativity, we see that B_ϵ is bimultiplicative, i.e.,

$$B_\epsilon(\alpha + \gamma, \beta) = B_\epsilon(\alpha, \beta)B_\epsilon(\gamma, \beta)\,, \qquad B_\epsilon(\beta, \alpha + \gamma) = B_\epsilon(\beta, \alpha)B_\epsilon(\beta, \gamma)\,.$$

Replacing e^α by $\epsilon_\alpha e^\alpha$ ($\epsilon_\alpha \in \mathbb{C}^\times, \epsilon_0 = 1$) gives an equivalent cocycle $\epsilon_1(\alpha, \beta) = \epsilon_\alpha \epsilon_\beta \epsilon_{\alpha+\beta}^{-1} \epsilon(\alpha, \beta)$ and does not change B_ϵ. We thus obtain a homomorphism, which we denote by b, from the second cohomology group $H^2(Q, \mathbb{C}^\times)$ ($=$ the group of equivalence classes of 2-cocycles) to the group of bimultiplicative skewsymmetric functions on $Q \times Q$ with respect to multiplication. The following lemma is well known.

LEMMA 5.5. *The homomorphism b is an isomorphism.*

PROOF. One has an obvious bijection between $H^2(Q, \mathbb{C}^\times)$ and isomorphism classes of central extensions $\widetilde{Q}$ of Q by $\mathbb{C}^\times$:

$$1 \to \mathbb{C}^\times \to \widetilde{Q} \to Q \to 1 .$$

Namely one defines on the set $\widetilde{Q} = Q \times \mathbb{C}$ a group multiplication by $(\alpha, \lambda)(\beta, \mu) = (\alpha + \beta, \epsilon(\alpha, \beta)\lambda\mu)$. Choosing a basis $\alpha_1, \dots, \alpha_\ell$ of Q, we may rearrange the product of two products of elements of the form $(\pm\alpha_i, \lambda)$, where indices are arranged in a non-decreasing order, such that the indices are non-decreasing by making use of the formula

$$(\alpha_i, \lambda)(\alpha_j, \mu)(\alpha_i, \lambda)^{-1} = (\alpha_j, B_\epsilon(\alpha_i, \alpha_j)\mu) .$$

This proves that b is bijective. $\square$

COROLLARY 5.5. *For any integral lattice Q there exists a unique up to equivalence solution $\epsilon(\alpha, \beta)$ to the equations (5.4.14a–c) taking only nonzero values. In this case we have:*

$$(5.5.4) \qquad p(\alpha) = (\alpha|\alpha) \mod 2 ,$$

so that

$$(5.5.5) \qquad B_\epsilon(\alpha, \beta) = (-1)^{(\alpha|\beta)+(\alpha|\alpha)(\beta|\beta)} .$$

PROOF. Letting $\alpha = \beta$ in (5.4.14b) we get (5.5.4). Since the function defined by (5.5.5) is bimultiplicative, the corollary follows now from Lemma 5.5. $\square$

Now we can prove the main result of this section.

THEOREM 5.5. *(a) Let Q be an integral lattice and let $V_Q = S \otimes \mathbb{C}_\epsilon[Q]$. Then there exists a simple vertex algebra structure on the space V_Q with the vacuum vector $|0\rangle = 1 \otimes 1$ and such that*

$$Y\left(\left(ht^{-1}\right) \otimes 1, z\right) = h(z), \quad h \in \mathfrak{h},$$

iff the bilinear form $(.|.)$ is non-degenerate. Such a vertex algebra structure is unique and is independent of the choice of the cocycle ϵ (satisfying (5.5.5)) up to isomorphism.

(b) The lattice vertex algebra described in (a) can be constructed as follows: Let $\epsilon : Q \times Q \to \{\pm 1\}$ be a 2-cocycle (i.e., (5.4.14a and c hold) such that

(5.5.6) $$\epsilon(\alpha, \beta)\epsilon(\beta, \alpha) = (-1)^{(\alpha|\beta)+(\alpha|\alpha)(\beta|\beta)}, \quad \alpha, \beta \in Q.$$

Consider the corresponding twisted group algebra $\mathbb{C}_\epsilon[Q]$ and the algebra $V_Q = S \otimes \mathbb{C}_\epsilon[Q]$. Then V_Q is the space of states with parity

(5.5.7) $$p(s \otimes e^\alpha) \equiv (\alpha|\alpha) \mod 2,$$

with the vacuum vector $|0\rangle = 1 \otimes 1$ and the infinitesimal translation operator T defined as the derivation of the algebra V_Q given by $(n > 0, h \in \mathfrak{h}, \alpha \in Q)$:

(5.5.8) $$T\left(\left(ht^{-n}\right) \otimes 1\right) = n\left(ht^{-n-1}\right) \otimes 1, \qquad T\left(1 \otimes e^\alpha\right) = \left(\alpha t^{-1}\right) \otimes e^\alpha.$$

For $\alpha \in Q$ let

(5.5.9) $$\Gamma_\alpha(z) = e^\alpha z^{\alpha_0} e^{-\sum_{j<0} \frac{z^{-j}}{j}\alpha_j} e^{-\sum_{j>0} \frac{z^{-j}}{j}\alpha_j},$$

where e^α is the operator of left multiplication by $1 \otimes e^\alpha$. Then the state-field correspondence is given by $(n_i \in \mathbb{Z}_+, h_i \in \mathfrak{h}, \alpha \in Q)$:

(5.5.10)
$$Y\left(\left(h_1 t^{-n_1-1}\right)\left(h_2 t^{-n_2-1}\right)\ldots \otimes e^\alpha, z\right) =: \partial^{(n_1)} h_1(z)\partial^{(n_2)} h_2(z)\ldots\Gamma_\alpha(z) : .$$

PROOF. If $h \in \mathfrak{h}$ is in the kernel of the bilinear form $(.|.)$, then $t^{-1}h \otimes 1$ generates an ideal of V_Q. Suppose that the bilinear form $(.|.)$ is non-degenerate. Then all solutions of the equations (5.4.14a–c) are of the form (5.4.13) (since $[h_0, c_\alpha] = 0$ for all $h \in \mathfrak{h}$). Since $1 \otimes J_\epsilon$ generates an ideal of V_Q (see (5.5.1)), it is necessary for simplicity of V_Q that $\epsilon(\alpha, \beta) \neq 0$ for all $\alpha, \beta \in Q$ (due to (5.5.2)). Due to Corollary 5.5, the structure of a vertex algebra on V_Q is unique up to isomorphism.

In order to complete the proof of (a), we must show that if the bilinear form $(.|.)$ is non-degenerate and $\epsilon(\alpha, \beta) \neq 0$ for all $\alpha\beta \in Q$, then V_Q is a simple vertex algebra. Recall that S is irreducible under all the operators h_n, $n \neq 0$, and that $h_0(1 \otimes e^\alpha) = (\alpha|h)1 \otimes e^\alpha$. It follows that any nonzero invariant subspace U of V_Q contains a vector $1 \otimes e^\alpha$ for some $\alpha \in Q$. Applying $z^{(\alpha|\alpha)}\Gamma_{-\alpha}(z)$ to this vector and letting $z = 0$, we conclude that $|0\rangle \in U$ (since $\epsilon(\alpha, -\alpha) \neq 0$), hence $U = V_Q$.

(b) is a reformulation of a special case of Proposition 5.4. $\qquad\square$

REMARK 5.5a. For an integral lattice Q one can construct explicitly a cocycle $\epsilon(\alpha, \beta)$ with values ± 1 such that

$$(5.5.11) \qquad \epsilon(\alpha, \beta)\epsilon(\beta, \alpha) = B(\alpha, \beta), \qquad \text{where } B(\alpha, \beta) = (-1)^{(\alpha|\beta)+(\alpha|\alpha)(\beta|\beta)},$$

as follows. Choose an ordered basis $\alpha_1, \ldots, \alpha_\ell$ of Q over $\mathbb{Z}$, let

$$\epsilon(\alpha_i, \alpha_j) = \begin{cases} B(\alpha_i, \alpha_j) & \text{if } i < j, \\ (-1)^{((\alpha_i|\alpha_i)+(\alpha_i|\alpha_i)^2)/2} & \text{if } i = j, \\ 1 & \text{if } i > j, \end{cases}$$

and extend to Q by bimultiplicativity. We thus obtain a bimultiplicative function $\epsilon : Q \times Q \to \{\pm 1\}$ such that

$$(5.5.12) \qquad \epsilon(\alpha, \alpha) = (-1)^{((\alpha|\alpha)+(\alpha|\alpha)^2)/2}, \quad \alpha \in Q.$$

Then bimultiplicativity implies the cocycle properties (i.e., (5.4.14a and c)), and (5.5.12) along with the bimultiplicativity imply (5.5.11).

The operators $\Gamma_\alpha(z)$ go back to the early days of string theory under the name vertex operators (see Remark 5.2). The only essential missing ingredient was the cocycle $\epsilon(\alpha, \beta)$ which was introduced by [**FK**].

EXAMPLE 5.5a. The main result of Section 5.2 states that the vertex algebra F of charged free fermions is isomorphic to the lattice vertex algebra $V_{\mathbb{Z}}$, where $\mathbb{Z}$ is the 1-dimensional lattice with the bilinear form $(m|n) = mn$. In this case $p(m) \equiv m \mod 2$ and $B_\epsilon(m, n) = 1$ for all $m, n \in \mathbb{Z}$, so that one may take $\epsilon(\alpha, \beta) = 1$ for all $\alpha, \beta \in \mathbb{Z}$.

More generally, let $Q = \mathbb{Z}^\ell$ with the standard bilinear form $(e_i|e_j) = \delta_{ij}$ where $\{e_i\}$ is the standard basis of $\mathbb{Z}^\ell$. Define a bimultiplicative function ϵ on $Q \times Q$ with values ± 1 by letting

$$\epsilon(e_i, e_j) = \begin{cases} 1 & \text{if } i \le j, \\ -1 & \text{if } i > j. \end{cases}$$

Then ϵ satisfies (5.5.12), hence satisfies equations (5.4.14a–c) (with $p(\alpha)$ defined by (5.5.4)). The corresponding lattice vertex algebra is isomorphic to $F^{\otimes \ell}$.

REMARK 5.5b. Let Q be the orthogonal direct sum of lattices Q_1 and Q_2 and assume that the 2-cocycle ϵ takes only nonzero values. Then we have an isomorphism of the (uniquely) associated vertex algebras: $V_Q \simeq V_{Q_1} \otimes V_{Q_2}$. In particular, let $\alpha \in Q$ be such that $(\alpha|\alpha) = 1$, so that we have a direct sum of lattices $Q = \mathbb{Z}\alpha \oplus \alpha^\perp$ and an isomorphism of vertex algebras:

$$V_Q = F \otimes V_{\alpha^\perp} .$$

It is convenient to collect together the most important properties of the fields $h(z)$ ($h \in \mathfrak{h}$) and the fields (vertex operators) $\Gamma_\alpha(z)$ defined by (5.5.9):

$$(5.5.13) \qquad h(z)h'(w) \sim \frac{(h|h')}{(z-w)^2} \qquad\qquad (h, h' \in \mathfrak{h}),$$

$$(5.5.14) \qquad h(z)\Gamma_\alpha(w) \sim \frac{(\alpha|h)\Gamma_\alpha(w)}{z-w} \qquad\qquad (h \in \mathfrak{h}, \alpha \in Q),$$

$$(5.5.15) \qquad \Gamma_\alpha(z)\Gamma_\beta(w) \sim \epsilon(\alpha,\beta)(z-w)^{(\alpha|\beta)}\Gamma_{\alpha,\beta}(z,w) \qquad (\alpha, \beta \in Q),$$

where $\Gamma_{\alpha,\beta}(z, w)$ is the following field in z and w:

$$\Gamma_{\alpha,\beta}(z,w) = e^{\alpha+\beta} z^{\alpha_0} w^{\beta_0} e^{-\sum_{j<0}\left(\frac{z^{-j}}{j}\alpha_j + \frac{w^{-j}}{j}\beta_j\right)} e^{-\sum_{j>0}\left(\frac{z^{-j}}{j}\alpha_j + \frac{w^{-j}}{j}\beta_j\right)} ,$$

$$(5.5.16) \qquad\qquad \partial\Gamma_\alpha(z) =: \alpha(z)\Gamma_\alpha(z): \qquad (\alpha \in Q).$$

These are equations (5.4.1b), (5.4.3a), (5.4.5a–b), and (5.4.9) respectively. It is straightforward to check that (cf. (5.5.16)):

$$\partial_z\Gamma_{\alpha,\beta}(z,w) =: \alpha(z)\Gamma_{\alpha,\beta}(z,w): \qquad (\equiv \alpha(z)_+\Gamma_{\alpha,\beta}(z,w) + \Gamma_{\alpha,\beta}(z,w)\alpha(z)_-).$$

By induction on n we obtain a formula for n-th derivative:

(5.5.17)
$$\partial_z^n \Gamma_{\alpha,\beta}(z,w) = \sum_{\substack{k_1+2k_2+\ldots=n \\ k_i \in \mathbb{Z}_+}} c_n(k_1, k_2, \ldots) : \alpha(z)^{k_1} \left(\partial\alpha(z)\right)^{k_2} \ldots \Gamma_{\alpha,\beta}(z,w) :,$$

where

$$c_n(k_1, k_2, \ldots) = \frac{n!}{(1!)^{k_1} k_1! (2!)^{k_2} k_2! \ldots}.$$

This is the number of partitions of n which contain k_i parts equal i. Expanding $\Gamma_{\alpha,\beta}(z,w)$ in a Taylor series by Lemma 3.1 and using (5.5.17) we obtain from (5.5.15) the following explicit OPE (up to an arbitrary order):

(5.5.18)
$$\Gamma_\alpha(z)\Gamma_\beta(w) \sim \epsilon(\alpha,\beta)(z-w)^{(\alpha|\beta)} \sum_{n \in \mathbb{Z}_+}$$

$$\sum_{\substack{k_1+2k_2+\ldots=n \\ k_i \in \mathbb{Z}_+}} \frac{(z-w)^n}{(1!)^{k_1} k_1! (2!)^{k_2} k_2! \ldots} : \alpha(w)^{k_1} \left(\partial\alpha(w)\right)^{k_2} \ldots \Gamma_{\alpha+\beta}(w) : .$$

Finally, we discuss the conformal structure of V_Q.

PROPOSITION 5.5. *Let Q be an integral lattice of rank ℓ and assume that the bilinear form $(.|.)$ is non-degenerate. Choose bases $\{a^i\}$ and $\{b^i\}$ of $\mathfrak{h}$ such that $(a^i|b^j) = \delta_{ij}$. Then*

(a) The vector

(5.5.19)
$$\nu = \frac{1}{2} \sum_{i=1}^{\ell} a^i_{-1} b^i_{-1} |0\rangle$$

is a conformal vector of the lattice vertex algebra V_Q (it is clearly independent of the choice of dual bases). The central charge of the corresponding Virasoro field $Y(\nu, z)$ is ℓ.

(b) The fields $h(z)$ ($h \in \mathfrak{h}$) are primary of conformal weight 1.

(c) The fields $\Gamma_\alpha(z) = Y(1 \otimes e^\alpha, z)$ are primary of conformal weight $\frac{1}{2}(\alpha|\alpha)$.

PROOF. It is straightforward to show, using Wick's theorem as in Section 3.5, that $Y(\nu, z) = \sum_{n \in \mathbb{Z}} L_n z^{-n-2}$ is a Virasoro field with central charge ℓ and that the $h(z)$ are primary fields of conformal weight 1 with respect to this Virasoro field for each $h \in \mathfrak{h}$. In particular,

(5.5.20)
$$L_{-1}(ht^{-1} \otimes 1) = ht^{-2} \otimes 1.$$

Since

$$Y(\nu, z) = \frac{1}{2} \sum_{i=1}^{\ell} : a^i(z) b^i(z) : ,$$

we have:

$$L_0 = \frac{1}{2} \sum_{i=1}^{\ell} a_0^i b_0^i + \frac{1}{2} \sum_{i=1}^{\ell} \sum_{n>0} \left(a_{-n}^i b_n^i + b_{-n}^i a_n^i \right) ,$$

$$L_{-1} = \frac{1}{2} \sum_{i=1}^{\ell} \sum_{n \geq 0} \left(a_{-n-1}^i b_n^i + b_{-n-1}^i a_n^i \right) ,$$

$$L_n = \frac{1}{2} \sum_{i=1}^{\ell} \sum_{j \in \mathbb{Z}} a_j^i b_{n-j}^i \quad \text{if } n \neq 0.$$

It follows that

(5.5.21)

$$L_0 \left((h_1 t^{-j_1})(h_2 t^{-j_2}) \ldots \otimes e^\alpha \right) = \left((j_1 + j_2 + \cdots) + \frac{1}{2}(\alpha|\alpha) \right) (h_1 t^{-j_1}) \ldots \otimes e^\alpha ,$$

(5.5.22)
$$L_{-1}(1 \otimes e^\alpha) \;=\; (\alpha t^{-1}) \otimes e^\alpha ,$$

(5.5.23)
$$L_n(1 \otimes e^\alpha) \;=\; 0 \qquad \text{for } n \geq 1.$$

Comparing (5.5.22) and (5.4.8), we see

(5.5.24)
$$L_{-1}(1 \otimes e^\alpha) \;=\; T(1 \otimes e^\alpha), \qquad \alpha \in Q .$$

Using the commutator formula (4.6.2a), formula (4.4.7), and (5.5.21–5.5.23), we see that $\Gamma_\alpha(z)$ is primary with respect to the Virasoro field $Y(\nu, z)$ of conformal weight $\frac{1}{2}(\alpha|\alpha)$.

In order to complete the proof of the proposition, it suffices to show that $L_{-1} = T$. But this follows from (5.5.20), (5.5.24) and Corollary 4.6g. $\qquad \square$

EXAMPLE 5.5b. Under the isomorphism $V_{\mathbb{Z}} \xrightarrow{\sim} F$, the conformal vector ν defined by (5.5.19) maps to the conformal vector $\nu^{1/2}$ (see (5.1.1)). Hence lattice vertex algebras may have several conformal structures.

5.6. Root lattice vertex algebras and affine vertex algebras

Let Q be a positive definite integral lattice. The set

$$\Delta = \{\alpha \in Q \,|\, (\alpha|\alpha) = 2\}$$

is called the *root system* for Q. It is well known (and easy to show, see e.g., [K2]) that Δ is isomorphic to a direct sum of finite root systems of type A, D and E. The lattice Q is called a *root lattice* if it is spanned over $\mathbb{Z}$ by the set Δ.

REMARK 5.6. The lattice Q is an orthogonal direct sum of the lattice $\mathbb{Z}^d$, where $\mathbb{Z}$ is the standard lattice of rank 1 and $d \geq 0$, and the sublattice $Q_{\geq 2} \subset Q$ spanned over $\mathbb{Z}$ by all α such that $(\alpha|\alpha) \geq 2$. Hence we have an isomorphism of the corresponding simple vertex algebras:

$$V_Q \simeq F^d \otimes V_{Q_{\geq 2}} \, .$$

In this section we will study the simple vertex algebra V_Q, where Q is a root lattice. We may assume that $\epsilon : Q \times Q \to \{\pm 1\}$ is a 2-cocycle (i.e., (5.4.14a and c) hold) such that

$$\epsilon(\alpha,\beta)\epsilon(\beta,\alpha) = (-1)^{(\alpha|\beta)} \, . \tag{5.6.1}$$

(cf. (5.5.6) and note that Q is an even lattice.)

Consider the generating fields $h(z) = \sum_{n\in\mathbb{Z}} h_n z^{-n-1}$ ($h \in \mathfrak{h} = \mathbb{C} \otimes_{\mathbb{Z}} Q$) and $\Gamma_\alpha(z) = \sum_{n\in\mathbb{Z}} e_n^\alpha z^{-n-1}$ ($\alpha \in \Delta$). Note that since Q is integral and positive definite we have only the following three possibilities for a pair $\alpha, \beta \in \Delta$:

$$(\alpha|\beta) \geq 0, \qquad (\alpha|\beta) = -1, \qquad or \qquad \alpha = -\beta \, .$$

Hence the following is a complete list of the OPE between the generating fields (see (5.5.13), (5.5.14) and (5.5.18)):

$$h(z)h'(w) \sim \frac{(h|h')}{(z-w)^2} \qquad\qquad \text{if } h, h' \in \mathfrak{h}, \tag{5.6.2a}$$

$$h(z)\Gamma_\alpha(w) \sim \frac{(\alpha|h)\Gamma_\alpha(w)}{z-w} \qquad\qquad \text{if } h \in \mathfrak{h},\ \alpha \in \Delta, \tag{5.6.2b}$$

$$\Gamma_\alpha(z)\Gamma_\beta(w) \sim 0 \qquad\qquad \text{if } \alpha, \beta \in \Delta,\ (\alpha|\beta) \geq 0. \tag{5.6.2c}$$

$$\Gamma_\alpha(z)\Gamma_\beta(w) \sim \epsilon(\alpha,\beta)\frac{\Gamma_{\alpha+\beta}(w)}{z-w} \qquad\qquad \text{if } \alpha, \beta \in \Delta,\ (\alpha|\beta) = -1, \tag{5.6.2d}$$

$$\Gamma_\alpha(z)\Gamma_{-\alpha}(w) \sim \frac{\epsilon(\alpha,-\alpha)}{(z-w)^2} + \frac{\epsilon(\alpha,-\alpha)\alpha(w)}{z-w} \quad \text{if } \alpha \in \Delta. \tag{5.6.2e}$$

These OPE are equivalent to the following commutation relations respectively $(m, n \in \mathbb{Z})$:

$$(5.6.3a) \qquad [h_m, h'_n] = m\delta_{m,-n}(h|h') \qquad\qquad \text{if } h, h' \in \mathfrak{h},$$

$$(5.6.3b) \qquad [h_m, e_n^\alpha] = (h|\alpha)e_{m+n}^\alpha \qquad\qquad \text{if } h \in \mathfrak{h}, \ \alpha \in \Delta,$$

$$(5.6.3c) \qquad [e_m^\alpha, e_n^\beta] = 0 \qquad\qquad \text{if } \alpha, \beta \in \Delta, \ (\alpha|\beta) \geq 0,$$

$$(5.6.3d) \qquad [e_m^\alpha, e_n^\beta] = \epsilon(\alpha, \beta)e_{m+n}^{\alpha+\beta} \qquad\qquad \text{if } \alpha, \beta \in \Delta, \ (\alpha|\beta) = -1,$$

$$(5.6.3e) \qquad [e_m^\alpha, e_n^{-\alpha}] = \epsilon(\alpha, -\alpha)\left(\alpha_{m+n} + m\delta_{m,-n}\right) \quad \text{if } \alpha \in \Delta.$$

Commutation relations (5.6.3a–e) lead us to consider the vector space

$$(5.6.4) \qquad\qquad \mathfrak{g} = \mathfrak{h} \oplus \left(\oplus_{\alpha\in\Delta}\mathbb{C}e_\alpha\right)$$

with the bracket defined by

$$(5.6.5a) \qquad\qquad [h, h'] = 0 \qquad\qquad \text{if } h, h' \in \mathfrak{h},$$

$$(5.6.5b) \qquad\qquad [h, e_\alpha] = (h|\alpha)e_\alpha \qquad\qquad \text{if } h \in \mathfrak{h}, \ \alpha \in \Delta,$$

$$(5.6.5c) \qquad\qquad [e_\alpha, e_\beta] = 0 \qquad\qquad \text{if } \alpha, \beta \in \Delta, \ (\alpha|\beta) \geq 0,$$

$$(5.6.5d) \qquad\qquad [e_\alpha, e_\beta] = \epsilon(\alpha, \beta)e_{\alpha+\beta} \qquad\qquad \text{if } \alpha, \beta \in \Delta, \ (\alpha|\beta) = -1,$$

$$(5.6.5e) \qquad\qquad [e_\alpha, e_{-\alpha}] = \epsilon(\alpha, -\alpha)\alpha \qquad\qquad \text{if } \alpha \in \Delta,$$

and with the $\mathbb{C}$-valued symmetric bilinear form $(.|.) : \mathfrak{g} \times \mathfrak{g} \to \mathbb{C}$ which extends that on $\mathfrak{h}$ by letting

$$(5.6.6) \qquad (e_\alpha|e_{-\alpha}) = \epsilon(\alpha, -\alpha), \qquad (e_\alpha|e_\beta) = 0 \ \text{ if } \alpha \neq -\beta, \qquad (\mathfrak{h}|e_\alpha) = 0.$$

We arrive at the following theorem, which is usually referred to as the *Frenkel-Kac construction* [**FK**].

THEOREM 5.6. *(a) The space* $\mathfrak{g}$ *with the bracket defined by (5.6.5a–e) is a semisimple Lie algebra with a Cartan subalgebra* $\mathfrak{h}$ *and the root space decomposition (5.6.4). The form* $(.|.)$ *is the non-degenerate symmetric invariant bilinear form on* $\mathfrak{g}$ *normalized by the condition* $(\alpha|\alpha) = 2$ *for* $\alpha \in \Delta$.

(b) Formulas (5.6.3a–e) define an irreducible representation of the affinization
$\widehat{\mathfrak{g}} = \mathfrak{g}[t, t^{-1}] + \mathbb{C}K$ *of the pair* $(\mathfrak{g}, (.|.))$ *with central charge* $k = 1$ *and highest weight*
vector $|0\rangle$ *such that*

$$(5.6.7) \qquad\qquad \mathfrak{g}[t]|0\rangle = 0 \,.$$

(c) The simple vertex algebra V_Q *is isomorphic to the affine vertex algebra*
$V^1(\mathfrak{g})$.

PROOF. The fact that $\mathfrak{g}$ is a Lie algebra follows from (5.6.3a–e) with $m = n = 0$.
Formulas (5.6.3a–e) also define a representation of the affinization of $(\mathfrak{g}, (.|.))$ with
$k = 1$ in the space V_Q (cf. (2.5.4)). It follows that the form $(.|.)$ is invariant, and it
is clearly symmetric and non-degenerate. It is also clear that $\mathfrak{g}$ is a semi-simple Lie
algebra. Furthermore, since $h(z)$ and $\Gamma_\alpha(z)$ are generating fields, it follows from
Theorem 5.5a that the representation of $\widehat{\mathfrak{g}}$ in V_Q defined by (5.6.3a–e) is irreducible.
Formula (5.6.7) holds since $e_n^\alpha|0\rangle = 0$ for $n \geq 0$. This completes the proof of the
theorem. $\qquad\qquad\qquad\qquad\qquad\qquad\qquad\qquad\qquad\qquad\qquad\qquad\square$

5.7. Conformal structure for affine vertex algebras

Let $\mathfrak{g}$ be a finite-dimensional Lie superalgebra with a supersymmetric invariant
bilinear form $(.|.)$, and let $\widehat{\mathfrak{g}} = \mathfrak{g}[t, t^{-1}] + \mathbb{C}K$ be the associated affinization (see Sec-
tion 2.5). Given $k \in \mathbb{C}$, consider the affine vertex algebra $V^k(\mathfrak{g})$ (see Example 4.9b)
graded by the Hamiltonian H $(= -t\partial_t)$. Recall that by Example 4.9b we have:

$$(5.7.1) \qquad\qquad \mathrm{Vac}\, V^k(\widehat{\mathfrak{g}}) = \mathbb{C}|0\rangle \,.$$

REMARK 5.7a. Due to Remark 4.9c, for any $a \in \mathfrak{g}_{\bar{0}}$, the exponential $e^{a_{(0)}}$ converges to an automorphism of the vertex algebra $V^k(\widehat{\mathfrak{g}})$. All these automorphisms
generate a group called the *group of inner automorphisms* of $V^k(\widehat{\mathfrak{g}})$.

In the previous section we established an isomorphism of $V^1(\widehat{\mathfrak{g}})$ with the root
lattice vertex algebra in the case when $\mathfrak{g}$ is a semi-simple Lie algebra and $(\alpha|\alpha) = 2$
for all roots α. This provides $V^1(\widehat{\mathfrak{g}})$ with a conformal structure (constructed in
Section 5.5).

In this section we give a construction of a conformal structure, which goes
back to Sugawara, for any (universal) affine vertex algebra in the case when $\mathfrak{g}$ is

an arbitrary simple (or commutative) Lie superalgebra, the bilinear form $(.|.)$ is non-degenerate, and k is different from a certain "critical" value. The construction gives the same vector as in the above mentioned special case under the isomorphism given by Theorem 5.6.

Note that for ν to be a conformal vector it suffices that

$$(5.7.2a) \qquad H\nu = 2\nu\,,$$

$$(5.7.2b) \qquad Y(\nu, z)g(w) \sim \frac{\partial g(w)}{z - w} + \frac{g(w)}{(z - w)^2} + \cdots \qquad \text{for all } g \in \mathfrak{g},$$

where $g(z) = \sum_n g_n z^{-n-1}$. Indeed, letting $Y(\nu, z) = \sum_{n \in \mathbb{Z}} L_n z^{-n-2}$, by the commutator formula (4.6.4), formula (5.7.2b) gives:

$$[L_{-1}, g(z)] = \partial g(z)\,,$$

$$[L_0, g_{-1}] = g_{-1}\,.$$

Since the fields $g(z)$, $g \in \mathfrak{g}$, are generating fields, it follows that

$$(5.7.3) \qquad L_{-1} = T\,, \qquad L_0 = H\,.$$

Since all eigenvalues of H are non-negative integers and the zero eigenspace is $\mathbb{C}|0\rangle$, we see from (5.7.3) and (5.7.2a) that $L_2\nu \in \mathbb{C}|0\rangle$ and $L_n\nu = 0$ for $n > 2$. Hence ν is a conformal vector by Theorem 4.10c.

By the general OPE formula (4.6.2a) we have for $g \in \mathfrak{g}$:

$$g(w)Y(\nu, z) \sim -\frac{Y(g_0\nu, z)}{z - w} + \frac{Y(g_1\nu, z)}{(z - w)^2} + \frac{Y(g_2\nu, z)}{(z - w)^3}\,.$$

Using Taylor's formula and (4.4.7), this becomes:

$$(5.7.4) \qquad g(w)Y(\nu, z) \sim \frac{Y(Tg_1\nu - g_0\nu, w)}{z - w} + \frac{Y(g_1\nu, w)}{(z - w)^2} + \frac{\alpha(g)}{(z - w)^3}\,,$$

where $\alpha(g) \in \mathbb{C}$ is defined by $g_2\nu = \alpha(g)|0\rangle$.

Comparing (5.7.2b) and (5.7.4) and using locality, we see that (5.7.2b) is equivalent to the following system of equations on $g \in \mathfrak{g}$:

$$(5.7.5) \qquad g_{-2}|0\rangle = Tg_1\nu - g_0\nu\,, \qquad g_{-1}|0\rangle = g_1\nu\,.$$

Applying T to both sides of the second equation, we get

$$g_{-2}|0\rangle = Tg_1\nu\,.$$

Substituting this in the first of the equations (5.7.5) and using (5.7.1), we arrive at the following statement.

PROPOSITION 5.7. *Let ν satisfy (5.7.2a). Then ν is a conformal vector if and only if the following equations hold for all $g \in \mathfrak{g}$:*

$$(5.7.6) \qquad g_0 \nu = 0, \qquad g_{-1}|0\rangle = g_1 \nu.$$

The field $g(z)$ is primary (of conformal weight 1) with respect to $Y(\nu, z)$ iff $\alpha(g) = 0$.

A vector ν satisfying (5.7.2a) can be written in the form:

$$\nu = \lambda \sum_i a^i_{-1} b^i_{-1}|0\rangle + d_{-2}|0\rangle,$$

for some a^i, b^i, $d \in \mathfrak{g}$; $\lambda \in \mathbb{C}$ is a parameter introduced for convenience. Equations (5.7.6) then turn into

$$(5.7.7a) \quad \lambda \sum_i \left([g, a^i]_{-1} b^i_{-1}|0\rangle + (-1)^{p(g)p(a^i)} a^i_{-1}[g, b^i]_{-1}|0\rangle \right) + [g, d]_{-2}|0\rangle = 0,$$

$$(5.7.7b) \quad g_{-1}|0\rangle = \lambda \left(\sum_i [[g, a^i], b^i]_{-1} + k(g|a^i) b^i_{-1} + (-1)^{p(g)p(a^i)} k(g|b^i) a^i_{-1} \right) |0\rangle$$
$$+ [g, d]_{-1}|0\rangle.$$

We also have

$$(5.7.8) \qquad \alpha(g) = \lambda k \sum_i \left([g, a^i]|b^i) \right) + 2k(g|d).$$

Suppose now that the bilinear form $(.|.)$ is non-degenerate and that $\{a^i\}$ and $\{b^i\}$ are dual bases of $\mathfrak{g}$, i.e., (3.5.3) holds. Let

$$\Omega = \sum_i a^i \otimes b^i \in \mathfrak{g} \otimes \mathfrak{g}$$

be the Casimir operator.

LEMMA 5.7. *(a) The element Ω is annihilated by the adjoint action of $\mathfrak{g}$ on $\mathfrak{g} \otimes \mathfrak{g}$.*

(b) The element $\bar{\Omega} := \sum_i a^i b^i \in U(\mathfrak{g})$ is central.

(c) If $\mathfrak{g}$ is simple or commutative, then

$$(5.7.9) \qquad \sum_i [a^i, [b^i, g]] = 2h^\vee g \qquad \text{for all } g \in \mathfrak{g},$$

where $2h^\vee$ is the eigenvalue of $\bar{\Omega}$ in the adjoint representation.

(d) The operators Ω and $\bar{\Omega}$ are independent of the choice of dual bases. In particular,

$$(5.7.10) \qquad\qquad \sum_i [a^i, b^i] = 0.$$

PROOF. We have by (3.5.4) and invariance of $(.|.)$:

$$[g, a^i] = \sum_j \left(b^j | [g, a^i] \right) a^j = \sum_j \left([b^j, g] | a^i \right) a^j,$$

$$[g, b^i] = \sum_j \left([g, b^i] | a^j \right) b^j = -\sum_j (-1)^{p(b^i)p(g)} \left([b^i, g] | a^j \right) b^j.$$

Hence

$$[g, \Omega] = \sum_i [g, a^i] \otimes b^i + \sum_i (-1)^{p(a_i)p(g)} a^i \otimes [g, b^i]$$

$$= \sum_{i,j} \left([b^j, g] | a^i \right) a^j \otimes b^i - \sum_{i,j} \left([b^i, g] | a^j \right) a^i \otimes b^j = 0,$$

proving (a).

(b) follows from (a) by considering the $\mathfrak{g}$-module homomorphism $\mathfrak{g} \otimes \mathfrak{g} \to U(\mathfrak{g})$ given by $x \otimes y \mapsto xy$.

If $\mathfrak{g}$ is simple, its adjoint representation is irreducible, hence, being a central element, $\bar{\Omega}$ acts on $\mathfrak{g}$ as a scalar, hence (c) holds. If $\mathfrak{g}$ is commutative, then (c) trivially holds with $h^\vee = 0$.

The independence of Ω and $\bar{\Omega}$ of the choice of dual bases is straightforward. Taking dual bases $\{b^i\}$ and $\{(-1)^{p(a^i)}a^i\}$, we deduce (5.7.10). This completes the proof. $\qquad\qquad\square$

The number $h^\vee$ is called the *dual Coxeter number* of the pair $(\mathfrak{g}, (.|.))$, where $\mathfrak{g}$ is a simple Lie superalgebra and $(.|.)$ is a non-degenerate invariant supersymmetric bilinear form. One usually normalizes the bilinear form $(.|.)$ by the condition that the maximal square length of a root equals 2. Then in the Lie algebra case, $h^\vee$ is always a positive integer listed, e.g., in [**K2**, Chapter 6]. (For the Lie superalgebra case see [**KW2**].)

Considering the $\mathfrak{g}$-module homomorphism $\mathfrak{g} \otimes \mathfrak{g} \to V^k(\mathfrak{g})$ given by $x \otimes y \mapsto x_{-1}y_{-1}|0\rangle$, we deduce from Lemma 5.7a that the sum in (5.7.7a) is zero. Hence equation (5.7.7a) holds, provided that the element d is central.

Suppose now that $\mathfrak{g}$ is simple or commutative and that $d \in \mathfrak{g}$ is central. Then, due to (3.5.4), (5.7.9) and (5.7.10), equation (5.7.7b) turns into the following simple equation:

$$g_{-1}|0\rangle = \lambda \left(2h^\vee + 2k\right) g_{-1}|0\rangle ,$$

which holds if we assume that $k \neq -h^\vee$ and let $\lambda = (2h^\vee + 2k)^{-1}$. Hence we proved that the element

$$\nu = \frac{1}{2(k + h^\vee)} \sum_i a^i_{(-1)} b^i_{(-1)} |0\rangle + d_{(-2)} |0\rangle$$

is a conformal vector of the vertex algebra $V^k(\mathfrak{g})$ (and of $\widetilde{V}^k(\mathfrak{g})$ too). In particular, the field

$$Y(\nu, z) = \frac{1}{2(k + h^\vee)} \sum_i a^i(z) b^i(z) + \partial d(z)$$

is a Virasoro field. This field is usually referred to as the *Sugawara construction* [S].

It follows from (5.7.8) that $\alpha(g) = 2k(g|d)$, hence all fields $g(z)$ are primary with respect to $Y(\nu, z)$ iff $d = 0$.

Recall that the case of commutative $\mathfrak{g}$ has been worked out in Section 4.9 using Wick's formula (see Proposition 4.9a). We state now the result in the case of simple $\mathfrak{g}$.

THEOREM 5.7. *Let $\mathfrak{g}$ be a simple finite-dimensional Lie superalgebra with a non-degenerate invariant supersymmetric bilinear form $(.|.)$ and let $\{a^i\}$ and $\{b^i\}$ be dual bases of $\mathfrak{g}$, i.e., $(b^i|a^j) = \delta_{ij}$. Then, provided that $k \neq -h^\vee$, where $h^\vee$ is defined by (5.7.9), the vector*

$$\nu = \frac{1}{2(k + h^\vee)} \sum_i a^i_{-1} b^i_{-1} |0\rangle$$

is a conformal vector of the vertex algebra $V^k(\mathfrak{g})$ (and $\widetilde{V}^k(\mathfrak{g})$) with central charge

$$(5.7.11) \qquad\qquad c_k = \frac{k \, \mathrm{sdim}\, \mathfrak{g}}{k + h^\vee} .$$

All fields $g(z)$, $g \in \mathfrak{g}$, are primary with respect to $Y(\nu, z)$ of conformal weight 1.

PROOF. It remains to calculate the central charge c of the Virasoro field $Y(\nu, z)$. We have for $g \in \mathfrak{g}$:

$$2(k + h^\vee)g_2\nu = \sum_i [g, a^i]_1 b^i_{-1}|0\rangle = k \sum_i \left([g, a^i]|b^i\right)|0\rangle$$

$$= k \sum_i \left(g|[a^i, b^i]\right)|0\rangle = 0$$

by (5.7.10). Note that $g_n\nu = 0$ if $n > 2$ for an obvious reason. Thus, recalling (5.7.6) we have for $g \in \mathfrak{g}$:

$$(5.7.12) \qquad g_n\nu = 0 \quad \text{for } n \geq 2 \text{ or } n = 0, \qquad g_1\nu = g_{-1}|0\rangle\,.$$

Next, we have:

$$L_2 = \frac{1}{2(k + h^\vee)} \sum_i \left(\sum_{n \leq -1} a^i_n b^i_{-n+2} + \sum_{n \geq 0}(-1)^{p(a^i)} b_{-n+2}a_n\right).$$

Using (5.7.12), we obtain:

$$2\left(k + h^\vee\right)L_2\nu = \sum_i (-1)^{p(a^i)} b^i_1 a^i_1 \nu = \sum_i (-1)^{p(a^i)} b^i_1 a^i_{-1}|0\rangle\,.$$

Hence

$$2(k + h^\vee)L_2\nu = k\sum_i (-1)^{p(a^i)}\left(b^i|a^i\right)|0\rangle = k\,\text{sdim}\,\mathfrak{g}|0\rangle\,,$$

proving (5.7.11). $\qquad\qquad\qquad\qquad\qquad\qquad\qquad\qquad\qquad\qquad\square$

REMARK 5.7b. A more straightforward (but somewhat less elegant) way to prove Theorem 5.7 is just to apply the non-commutative Wick formula (3.3.12) (cf. (4.7.8)).

REMARK 5.7c. Let $\mathfrak{g}$ be a finite-dimensional Lie superalgebra with a non-degenerate invariant supersymmetric bilinear form $(.|.)$, and suppose that $\mathfrak{g} = \oplus_{i=0}^n \mathfrak{g}^i$ is an (orthogonal) direct sum of a commutative subalgebra $\mathfrak{g}^0$ and simple subalgebras $\mathfrak{g}^i$, $i > 0$. Then

$$V^k(\widehat{\mathfrak{g}}) = \otimes_{i=0}^n V^{k_i}(\widehat{\mathfrak{g}^i})\,,$$

where $k = (k_0, k_1, \dots)$. Provided that $k_i \neq -h_i^\vee$ (note that $h_0^\vee = 0$) the vertex algebra $V^k(\widehat{\mathfrak{g}})$ is conformal with the conformal vector

$$\nu = \sum_{i=0}^n \nu^i\,,$$

where ν^i, the conformal vector of $V^{k_i}(\widehat{\mathfrak{g}^i})$ given by the Sugawara construction, is identified with $|0\rangle \otimes \cdots \otimes \nu^i \otimes \cdots \otimes |0\rangle$.

Due to Lemma 5.7b, the conformal vector ν is fixed by the group of inner automorphisms G of the vertex algebra $V^k(\widehat{\mathfrak{g}})$. Hence for any subgroup Γ of G the fixed point set subalgebra $V^k(\widehat{\mathfrak{g}})^\Gamma$ contains ν, hence is a conformal vertex algebra of the same rank c_k.

REMARK 5.7d. We constructed in Example 4.9b an operator T^* on $V^k(\widehat{\mathfrak{g}})$ for any $k \in \mathbb{C}$. It is easy to see that $T^* = L_1$ (if $k \neq -h^\vee$). Hence T^* satisfies (4.9.10) with $H = L_0$ for $k \neq -h^\vee$, hence for all k. Thus $V^k(\widehat{\mathfrak{g}})$ is Möbius conformal even at the critical value $k = -h^\vee$ (but is not conformal).

We turn now to the discussion of conformal structure for coset models. Let $\mathfrak{g}$ be a Lie superalgebra as in Remark 5.7c and let $\mathfrak{h}$ be a subalgebra of $\mathfrak{g}$ such that $(.|.)|_\mathfrak{h}$ is non-degenerate and $\mathfrak{h}$ is too a direct sum of simple and commutative subalgebras. Let $\nu_\mathfrak{g}$ and $\nu_\mathfrak{h}$ be the elements of $V^k(\widehat{\mathfrak{g}})$ given by the Sugawara construction, so that $Y(\nu_\mathfrak{g}, z) = \sum_n L_n^\mathfrak{g} z^{-n-2}$ and $Y(\nu_\mathfrak{h}, z) = \sum_n L_n^\mathfrak{h} z^{-n-2}$ are Virasoro fields. The fields $h(z)$ with $h \in \mathfrak{h}$ generate a vertex subalgebra of $V^k(\widehat{\mathfrak{g}})$ isomorphic to $V^{k'}(\mathfrak{h})$ and we denote by $C(k, \mathfrak{g}, \mathfrak{h})$ its centralizer (see Remark 4.6b). In other words (by Corollary 4.6 (b))

$$C(k, \mathfrak{g}, \mathfrak{h}) = \left\{ a \in V^k(\widehat{\mathfrak{g}}) \,|\, h_n(a) = 0 \quad \text{for all } h \in \mathfrak{h} \text{ and } n \in \mathbb{Z}_+ \right\}.$$

A conformal vector for the vertex algebra $C(k, \mathfrak{g}, \mathfrak{h})$ can be constructed as follows. This is known as the *Goddard-Kent-Olive construction* [**GKO**]. (Some further applications of this construction in representation theory may be found in [**KW1**] and [**KR**].)

COROLLARY 5.7. *The vector*

$$\nu = \nu_\mathfrak{g} - \nu_\mathfrak{h}$$

is a conformal vector of the vertex algebra $C(k, \mathfrak{g}, \mathfrak{h})$ with central charge equal the difference between central charges of $\nu_\mathfrak{g}$ and $\nu_\mathfrak{h}$.

PROOF. By Theorem 5.7 we have for all $h \in \mathfrak{h}$:

$$Y(\nu_\mathfrak{g}, z)h(w) \sim \frac{\partial h(w)}{z - w} + \frac{h(w)}{(z - w)^2} \sim Y(\nu_\mathfrak{h}, z)h(w).$$

It follows that

$$(5.7.13) \qquad\qquad Y(\nu, z)h(w) \sim 0 \qquad \text{for all } h \in \mathfrak{h}.$$

Hence $\nu \in C(k, \mathfrak{g}, \mathfrak{h})$. Let $Y(\nu, z) = \sum_n L_n z^{-n-2}$.

Next, by the construction, we see that $L^{\mathfrak{h}}_{-1}$ annihilates $C(k, \mathfrak{g}, \mathfrak{h})$, hence

$$(5.7.14) \qquad\qquad L_{-1} = L^{\mathfrak{g}}_{-1} = T \qquad\qquad \text{on } C(k, \mathfrak{g}, \mathfrak{h}).$$

Finally, $Y(\nu, z)$ is a Virasoro field since both $Y(\nu_{\mathfrak{g}}, z)$ and $Y(\nu_{\mathfrak{h}}, z)$ are Virasoro fields (by Theorem 5.7). Indeed, we have, using (5.7.13) twice:

$$Y(\nu, z)Y(\nu, w) \sim (Y(\nu_{\mathfrak{g}}, z) - Y(\nu_{\mathfrak{h}}, z))\, Y(\nu, w)$$

$$\sim Y(\nu_{\mathfrak{g}}, z)Y(\nu, w)$$

$$\sim Y(\nu_{\mathfrak{g}}, z)Y(\nu_{\mathfrak{g}}, w) - Y(\nu_{\mathfrak{g}}, z)Y(\nu_{\mathfrak{h}}, w)$$

$$\sim Y(\nu_{\mathfrak{g}}, z)Y(\nu_{\mathfrak{g}}, w) - (Y(\nu, z) + Y(\nu_{\mathfrak{h}}, z))\, Y(\nu_{\mathfrak{h}}, w)$$

$$\sim Y(\nu_{\mathfrak{g}}, z)Y(\nu_{\mathfrak{g}}, w) - Y(\nu_{\mathfrak{h}}, z)Y(\nu_{\mathfrak{h}}, w)\,.$$

$$\square$$

REMARK 5.7e. The vacuum subalgebra $\operatorname{Vac} C(k, \mathfrak{g}, \mathfrak{h})$ coincides with the zero eigenspace of L_0 and is often larger than $\mathbb{C}|0\rangle$. For example, if $\mathfrak{g}$ is a simple Lie algebra and $\mathfrak{h}$ is its conformal subalgebra (see e.g., [**K2**, Chapter 13]), then $\operatorname{Vac} C(1, \mathfrak{g}, \mathfrak{h}) = C(1, \mathfrak{g}, \mathfrak{h})$ is almost always larger than $|0\rangle$. Another example is $C(k, \mathfrak{g}, \mathfrak{h})$ where k is a positive integer, $\mathfrak{g}$ is a simply laced simple Lie algebra and $\mathfrak{h}$ is its Cartan subalgebra; then the vacuum subalgebra $\operatorname{Vac} C(k, \mathfrak{g}, \mathfrak{h})$ is isomorphic to the group algebra of the center of the simply connected Lie group with the Lie algebra $\mathfrak{g}$.

5.8. Super boson-fermion correspondence and sums of squares

Consider the $4 = 2 + 2$-dimensional superspace A with a basis $\varphi^+, \varphi^-, \psi^+, \psi^-$, where $\varphi^{\pm}$ are even elements and $\psi^{\pm}$ are odd elements. Define an anti-supersymmetric bilinear form $(.|.)$ on A by

$$(\varphi^-|\varphi^+) = 1,\ (\psi^-|\psi^+) = 1,\ (\varphi^{\pm}|\psi^{\pm}) = (\varphi^{\pm}|\psi^{\mp}) = (\psi^{\pm}|\psi^{\pm}) = 0\,.$$

Consider the Clifford affinization C_A of A. Recall (see Section 2.5 and 3.6) that this is a regular formal distribution Lie superalgebra spanned by coefficients of five pairwise local formal distributions: $\varphi^+(z)$, $\varphi^-(z)$, 1, which are even, and $\psi^+(z)$, $\psi^-(z)$, which are odd. All non-trivial OPEs are as follows:

$$(5.8.1) \qquad \varphi^+(z)\varphi^-(w) \sim -\frac{1}{z-w} \,,\; \varphi^-(z)\varphi^+(w) \sim \frac{1}{z-w} \,,$$

$$\psi^+(z)\psi^-(w) \sim \frac{1}{z-w} \,,\; \psi^-(z)\psi^+(w) \sim \frac{1}{z-w} \,.$$

It will be convenient to use the following expansions of these distributions:

$$\varphi^\pm(z) = \sum_{j\in\frac12+\mathbf{Z}} \varphi_j^\pm z^{-j-\frac12}\,,\; \psi^+(z) = \sum_{j\in\mathbf{Z}} \psi_j^+ z^{-j-1}\,,\; \psi^-(z) = \sum_{j\in\mathbf{Z}} \psi_j^- z^{-j}\,.$$

Then relations (5.8.1) are equivalent to the following commutation relation of the basis elements of the Lie superalgebra C_A:

$$(5.8.2) \qquad [\varphi_m^-, \varphi_n^+] = \delta_{m,-n}\,, [\varphi_m^\pm, \varphi_n^\pm] = 0\,, [\varphi_m^\pm, \psi_n^\pm] = 0\,,$$

$$[\varphi_m^\pm, \psi_n^\mp] = 0\,, [\psi_m^+, \psi_n^-] = \delta_{m,-n}\,, [\psi_m^\pm, \psi_n^\pm] = 0\,.$$

Recall that, by Theorem 3.6, the Lie superalgebra C_A has a unique irreducible module, which we shall denote by F_{super}, such that the central element 1 is represented by the identity operator and there exists a non-zero vector $|0\rangle$ such that

$$\varphi^\pm(z)_-|0\rangle = 0\,, \psi^\pm(z)_-|0\rangle = 0\,,$$

in other words:

$$(5.8.3) \qquad \varphi_j^\pm|0\rangle = 0 \text{ and } \psi_j^-|0\rangle = 0 \text{ for } j > 0,\; \psi_j^+|0\rangle = 0 \text{ for } j \geq 0\,.$$

The monomials

$$(5.8.4) \qquad \varphi_{-s_1}^- \cdots \varphi_{-s_{b^-}}^- \, \varphi_{-r_1}^+ \cdots \varphi_{-r_{b^+}}^+ \, \psi_{-j_1}^- \cdots \psi_{-j_{f^-}}^- \, \psi_{-i_1}^+ \cdots \psi_{-i_{f^+}}^+ |0\rangle$$

where $0 < s_1 \leq \ldots \leq s_{b^-}$, $0 < r_1 \leq \ldots \leq r_{b^+}$, $0 \leq j_1 < \ldots < j_{f^-}$ and $0 < i_1 < \ldots < i_{f^+}$ form a basis of F_{super}. The number $b = b^+ - b^-$ (resp. $f = f^+ - f^-$) is called the *bosonic* (resp. *fermionic*) *charge*, the number $b + f$ is called the *total charge*, and $-$(sum of all indices in (5.8.4)) is called the *energy* of the monomial (5.8.4). Denote by $F_j^{(b,f)}$ the linear span of all monomials (5.8.4) having bosonic charge b, fermionic charge f and energy j. Note that this is a finite-dimensional space (which is 0 unless $b, f \in \mathbb{Z}, j \in \frac{1}{2}\mathbb{Z}_+$). As in Section 5.1, we wish to compute the character:

$$chF_{\text{super}} = \sum_{b,f,j} (\dim F_j^{(b,f)}) q^j t^b s^f$$

in two different ways. Note that chF_{super} is a formal power series in $q^{1/2}$ with coefficients being Laurent polynomials in t and s.

As in Section 5.1, just looking at the basis (5.8.4), we obtain

$$(5.8.5) \qquad chF_{\text{super}} = \prod_{n=1}^{\infty} \frac{(1 + sq^n)(1 + s^{-1}q^{n-1})}{(1 - tq^{n-1/2})(1 - t^{-1}q^{n-1/2})}.$$

Here and further we shall assume that $|q| < |t|^2 < 1$ and $|q| < |s|^2 < 1$ (in order to be able to use expansions of geometric series that occur). Let $F_{\text{super}}^{(m)}$ denote the linear span of all monomials of total charge m. Then we have the total charge decomposition of F_{super} and the corresponding decomposition of characters:

$$(5.8.6) \qquad F_{\text{super}} = \oplus_{m\in\mathbb{Z}}F_{\text{super}}^{(m)}, \; chF_{\text{super}} = \sum_{m\in\mathbb{Z}} chF_{\text{super}}^{(m)}.$$

REMARK 5.8a. There exists a unique monomial of minimal energy in each $F_{\text{super}}^{(m)}$, which we shall denote by $|m\rangle$. It is as follows:

$$|m\rangle = (\varphi^+_{-\frac{1}{2}})^m|0\rangle \text{ if } m \geq 0, \; |m\rangle = (\varphi^-_{-\frac{1}{2}})^{-m-1}\psi_0^-|0\rangle \text{ if } m < 0.$$

We shall calculate each term $chF_{\text{super}}^{(m)}$ in (5.8.6) using *super-bosonization*. Let

$$\alpha(z) \;=\; :\psi^+(z)\psi^-(z):, \; \beta(z) =: \varphi^+(z)\varphi^-(z):,$$

$$E^-(z) \;=\; :\varphi^+(z)\psi^-(z):, \; E^+(z) =: \psi^+(z)\varphi^-(z): .$$

Using Wick's formula and (5.8.1), we easily obtain the OPEs for these fields:

$$(5.8.7a) \qquad \alpha(z)\alpha(w) \sim \frac{1}{(z-w)^2}, \; \beta(z)\beta(w) \sim -\frac{1}{(z-w)^2},$$

$$(5.8.7b) \qquad \alpha(z)E^\pm(w) \sim \frac{\pm E^\pm(w)}{z-w}, \; \beta(z)E^\pm(w) \sim \frac{\mp E^\pm(w)}{z-w},$$

$$(5.8.7c) \quad E^+(z)E^-(w) \sim \frac{\alpha(w) + \beta(w)}{z-w} + \frac{1}{(z-w)^2}, \; E^\pm(z)E^\pm(w) \sim 0.$$

We let

$$\alpha(z) = \sum_{n\in\mathbb{Z}} \alpha_n z^{-n-1}, \; \beta(z) = \sum_{n\in\mathbb{Z}} \beta_n z^{-n-1}.$$

REMARK 5.8b. The operator α_0 (resp. β_0) is the fermionic (resp. bosonic) charge operator (i.e., vectors (5.8.4) are eigenvectors for these operators whose eigenvalues are f and b, respectively). This is clear by (3.6.10). It is easy to show (cf. Section 3.6) that $L(z) =: \partial\psi^-(z)\psi^+(z) : +\frac{1}{2}(: \partial\varphi^+(z)\varphi^-(z) : - : \partial\varphi^-(z)\varphi^+(z) :)$ is a Virasoro field with central charge 1 such that L_0 is the energy operator and the fields $\psi^+(z)$, $\psi^-(z)$, $\varphi^\pm(z)$ have conformal weights 1, 0 and 1/2, respectively. Consequently, the fields $\alpha(z)$, $\beta(z)$, $E^+(z)$ and $E^-(z)$ have conformal weights 1, 1, 3/2 and 1/2, respectively. Hence (cf. Section 5.1):

$$ch F_{\text{super}} = \text{tr}_{F_{\text{super}}} q^{L_0} s^{\alpha_0} t^{\beta_0} .$$

Consider the Lie superalgebra $gl(1,1)$ with the invariant bilinear form $(A|B) = \text{str}\, AB$. (Recall that this is the Lie superalgebra defined by the bracket (2.3.1) on the $\mathbb{Z}/2\mathbb{Z}$-graded associative superalgebra of endomorphisms of the $1 + 1$-dimensional superspace, and that $\text{str} \begin{pmatrix} a & b \\ c & d \end{pmatrix} = a - d$.) Let $E^+ = \begin{pmatrix} 0 & 1 \\ 0 & 0 \end{pmatrix}$, $E^- = \begin{pmatrix} 0 & 0 \\ 1 & 0 \end{pmatrix}$, $\alpha = \begin{pmatrix} 1 & 0 \\ 0 & 0 \end{pmatrix}$, $\beta = \begin{pmatrix} 0 & 0 \\ 0 & 1 \end{pmatrix}$ be a basis of $gl(1,1)$ and let $\widehat{gl(1,1)}$ be the affinization of the pair $(gl(1,1), (.|.))$. Formulas (5.8.7a-c) show that associating to currents of the above four elements the fields $E^+(z)$, $E^-(z)$, $\alpha(z)$ and $\beta(z)$, respectively, defines a representation of the current algebra $\widehat{gl(1,1)}$ with central charge 1 in the space F_{super}, preserving the total charge decomposition (5.8.6).

Consider the lattice $Q = \mathbb{Z}\alpha + \mathbb{Z}\beta$ with the symmetric bilinear form $(.|.)$ coming from $gl(1,1)$, i.e.:

$$(\alpha|\alpha) = 1 , \ (\beta|\beta) = -1 , \ (\alpha|\beta) = 0 .$$

As in Section 5.2, we show that there exists a unique invertible operator e^α on F_{super} satisfying the following properties:

$$e^\alpha \psi_n^\pm = \psi_{n\mp1}^\pm e^\alpha , \ e^\alpha \varphi_n^\pm = \varphi_n^\pm e^\alpha , \ e^\alpha|0\rangle = \psi_{-1}^+|0\rangle .$$

As in Section 5.2, it is immediate to check

$$(5.8.8) \qquad [\alpha_n, e^\alpha] = \delta_{0,n} e^\alpha , \ [\beta_n, e^\alpha] = 0 , \ e^\alpha E_{(n)}^\pm = E_{(n\mp1)}^\pm e^\alpha .$$

Similarly, there exists a unique invertible operator e^β on F_{super} such that

$$e^\beta \psi_n^\pm = \psi_n^\pm e^\beta , \ e^\beta \varphi_n^\pm = \varphi_{n\mp1}^\pm e^\beta , \ e^\beta|0\rangle = \varphi_{-\frac{1}{2}}^-|0\rangle ,$$

and we have:

$$(5.8.9) \qquad [\alpha_n, e^\beta] = 0, \; [\beta_n, e^\beta] = -\delta_{0,n} e^\beta, \; e^\beta E^\pm_{(n)} = E^\pm_{(n\pm 1)} e^\beta.$$

Operators e^α and e^β commute. For any $\gamma = a\alpha + b\beta \in Q$ we let

$$\gamma_n = a\alpha_n + b\beta_n, \; \gamma(z) = a\alpha(z) + b\beta(z), \; e^\gamma = (e^\alpha)^a (e^\beta)^b.$$

Then we have (cf. (5.4.2)):

$$(5.8.10) \qquad [\gamma_n, e^\sigma] = \delta_{0,n}(\gamma|\sigma) e^\sigma, \quad \gamma, \sigma \in Q.$$

Introduce the following operators on F_{super} for any $\gamma \in Q$ (cf. Sections 5.2 and 5.5):

$$\Gamma^\pm_\gamma(z) = e^{\pm \sum_{j=1}^\infty \frac{z^{\pm j}}{j} \gamma_{\mp j}}, \; \Gamma_\gamma(z) = e^\gamma z^{\gamma_0} \Gamma^+_\gamma \Gamma^-_\gamma(z).$$

Recall that by the boson-fermion correspondence (see Section 5.2) we have:

$$(5.8.11) \qquad \psi^\pm(z) = \Gamma_{\pm\alpha}(z).$$

The next key formula is ([**KL1**]):

$$(5.8.12) \qquad \varphi^\pm(z) = z^{\pm\alpha_0} e^{\pm\alpha} \Gamma^+_{\pm\alpha}(z) E^\mp(z) \Gamma^-_{\pm\alpha}(z).$$

It is natural to call formulas (5.8.11) and (5.8.12) the *super boson-fermion correspondence*.

The proof of (5.8.12) uses the following relations:

$$(5.8.13a) \qquad E^+(z)\psi^-(w) \sim \frac{\varphi^-(w)}{z - w},$$

$$(5.8.13b) \qquad [\alpha_n, E^+(w)] = w^n E^+(w),$$

$$(5.8.13c) \qquad E^+(z)e^{-\alpha} = z^{-1} e^{-\alpha} E^+(z),$$

$$(5.8.13d) \qquad E^+(z)w^{-\alpha_0} = w^{-\alpha_0+1} E^+(z),$$

$$(5.8.13e) \qquad e^\alpha z^{\gamma_0} = z^{\gamma_0+(\alpha|\gamma)} e^\alpha, \quad \gamma \in Q.$$

Relation (5.8.13a) is immediate by Wick's formula, relation (5.8.13b) follows from (5.8.7b), relation (5.8.13c) and (5.8.13e) follow from (5.8.8) and relation (5.8.13d) follows from (5.8.13b) for $n = 0$.

Since $E^+(z)\psi^-(w) = E^+(z)e^{-\alpha}w^{-\alpha_0}\Gamma^+_{-\alpha}(w)\Gamma^-_{-\alpha}(w)$, in view of (5.8.13a) and (5.8.13c-e) in order to prove (5.8.12) for φ^-, it suffices to show that in the domain $|z| > |w|$ one has:

$$\Gamma^+_{-\alpha}(w)^{-1}E^+(z)\Gamma^+_{-\alpha}(w) = \frac{1}{1 - \frac{w}{z}}E^+(z).$$

But this follows from the simple fact that for any formal power series $a(w)$ one has:

$$(5.8.14) \qquad e^{a(w)}Ae^{-a(w)} = e^{ad\,a(w)}A,$$

applied to $a(w) = \sum_{n=1}^{\infty}\frac{w^n}{n}\alpha_{-n}$ and $A = E^+(z)$ by making use of (5.8.13b). The proof for φ^+ is similar.

The super boson-fermion correspondence implies the following important fact (similar to Theorem 5.1).

PROPOSITION 5.8. *The representation of the current algebra $\widehat{gl(1,1)}$ in each subspace $F_{\text{super}}^{(m)}$ is irreducible.*

PROOF. Let U be a non-zero $\widehat{gl(1,1)}$-invariant subspace of $F_{\text{super}}^{(m)}$. It follows from (5.8.8) that $e^{n\alpha}U$ is a $\widehat{gl(1,1)}$-invariant subspace of $F_{\text{super}}^{(m+n)}$. Hence $\widetilde{U} = \oplus_{n\in\mathbb{Z}}e^{n\alpha}U$ is a e^{α}-invariant and $\widehat{gl(1,1)}$-invariant subspace of F_{super}. Therefore, due to (5.8.11) and (5.8.12), $\widetilde{U}$ is invariant with respect to the whole algebra C_A. Due to irreducibility of F_{super}, $\widetilde{U} = F_{\text{super}}$ and hence $U = F_{\text{super}}^{(m)}$.

$\square$

Note that the the element $\delta := \alpha+\beta$ is a central element of the Lie superalgebra $gl(1,1)$ and $(\delta|\delta) = 0$. It follows from formulas (5.8.7b) and (5.8.8)-(5.8.10) that

$$(5.8.15) \qquad [\Gamma_{\pm\delta}(z), E^+(w)] = 0\,, \quad [\Gamma_{\pm\delta}(z), E^-(w)] = 0\,.$$

Also, we have (see e.g. (5.5.18)) the following OPE:

$$(5.8.16) \qquad \Gamma_{-\delta}(z)\Gamma_{\delta}(w) \sim (1 - \delta(w)(z - w) + \dots)\,.$$

Formulas (5.8.7c), (5.8.15) and (5.8.16) imply the following key formula:

$$(5.8.17) \qquad (\Gamma_{-\delta}(z)E^+(z))(\Gamma_{\delta}(w)E^-(w)) \sim \frac{1}{(z - w)^2}\,.$$

In order to understand better the representation of $gl(1,1)\hat{}$ in $F_{\text{super}}^{(m)}$, introduce the following odd fields:

$$\xi^{\pm}(z) = \Gamma_{\mp\delta}^{+}(z)\Gamma_{\mp\delta}^{-}(z)E^{\pm}(z)\,.$$

Since $\xi^{\pm}(z) = e^{\pm\delta}z^{\pm\delta_0}(\Gamma_{\mp\delta}(z)E^{\pm}(z))$, we get from (5.4.3b) and (5.8.13b) that

$$(5.8.18) \qquad\qquad [\alpha_n, \xi^{\pm}(w)] = \pm\delta_{n,0}\xi^{\pm}(w)\,,$$

and we obtain from (5.8.17) (using Taylor's formula):

$$\xi^{+}(z)\xi^{-}(w) \sim (\frac{z}{w})^{\delta_0}\frac{1}{(z-w)^2} \sim \frac{1}{(z-w)^2} + \frac{\delta_0 w^{-1}}{z-w}\,.$$

Obviously, we also have $\xi^{\pm}(z)\xi^{\pm}(w) \sim 0$. It is clear from the definition and Remark 5.8b that the conformal weights of the fields $\xi^{+}(z)$ and $\xi^{-}(z)$ are 3/2 and 1/2 respectively. Taking the series expansions

$$\xi^{+}(z) = \sum_{j\in\frac{1}{2}+\mathbf{Z}} \xi_j^{+} z^{-j-3/2}\,,\quad \xi^{-}(z) = \sum_{j\in\frac{1}{2}+\mathbf{Z}} \xi_j^{-} z^{-j-1/2}\,,$$

the last two OPEs are translated as follows on each $F_{\text{super}}^{(m)}$ (cf. [**Be**]):

$$(5.8.19) \qquad [\xi_r^{+}, \xi_s^{-}] = (r+m+\frac{1}{2})\delta_{r,-s}\,,\ [\xi_r^{\pm}, \xi_s^{\pm}] = 0\ (r,s \in \frac{1}{2}+\mathbb{Z})\,.$$

It is easy now to complete the proof of the following theorem.

THEOREM 5.8. *(a) The space $F_{\text{super}}^{(m)}$ is irreducible with respect to all operators α_n, β_n $(n \in \mathbb{Z})$ and ξ_r^{+}, ξ_r^{-} $(r \in \frac{1}{2}+\mathbb{Z})$.*

(b) The following elements form a basis of $F_{\text{super}}^{(m)}$:

$$\alpha_{-i_1} \cdots \ \beta_{-j_1} \cdots \ \xi_{-k_1}^{+} \cdots \ \xi_{-r_1}^{-} \cdots \ |m\rangle$$

where the i's and the j's are non-decreasing finite sequences of positive integers and the k's and the r's are increasing sequences of positive half-odd-integers such that all k_i are different from $m + \frac{1}{2}$ if $m \geq 0$ and all r_i are different from $-m - \frac{1}{2}$ if $m < 0$.

(c) The energy of the operators α_n, β_n (resp. $\xi_r^{\pm}$) is $-n$ (resp. $-r$) and their bosonic and fermionic charges are 0 (respectively ∓ 1 and ± 1).

(d) The characters of the $gl(1,1)\hat{\ }$-modules $F_{\text{super}}^{(m)}(m \in \mathbb{Z})$ are as follows:

$$ch F_{\text{super}}^{(m)} \; = \; \frac{t^m q^{\frac{1}{2}m}}{1 + s^{-1}q^{m+1/2}} D \;\; \text{if} \;\; m \geq 0,$$

$$ch F_{\text{super}}^{(m)} \; = \; \frac{s^{-1}t^{m+1}q^{-(m+1)/2}}{1 + s^{-1}tq^{-m-1/2}} D \;\; \text{if} \;\; m < 0,$$

where

$$D = \prod_{n=1}^{\infty} \frac{(1 + st^{-1}q^{n-1/2})(1 + s^{-1}tq^{n-1/2})}{(1 - q^n)^2}.$$

PROOF. By definition, the coefficients of the fields $E^{\pm}(z)$ can be expressed via the operators δ_n and $\xi_r^{\pm}$ (since the operators $\Gamma_{\gamma}^{\pm}$ are invertible). Hence the whole representation of $gl(1,1)\hat{\ }$ in $F_{\text{super}}^{(m)}$ can be expressed in terms of the operators α_n, β_n, $\xi_r^{\pm}$. Consequently, due to Proposition 5.8, these operators act irreducibly on $F_{\text{super}}^{(m)}$, proving (a). Statement (c) follows from definitions. Note that the operators $\xi_{-m-1/2}^{+}$ and $\xi_{m+\frac{1}{2}}^{-}$ commute with all operators $\alpha_n, \beta_n, \xi_r^{\pm}$ on $F_{\text{super}}^{(m)}$ (see (5.8.18) and (5.8.19)) and hence, due to (a) act as scalars. By (c) these scalars are 0. Due to (5.8.7a), (5.8.18) and (5.8.19), (b) follows from Theorems 3.5(a) and 3.6(a). Statement (d) is immediate from (b),(c) and Remark 5.8a.

$\square$

REMARK 5.8c. Inserting the expression for $E^{\pm}(z)$ from above in (5.8.12), we obtain:

$$\varphi^{\pm}(z) = z^{\pm\alpha_0} e^{\pm\alpha} \Gamma_{\mp\beta}^{+}(z)\xi^{\mp}(z)\Gamma_{\mp\beta}^{-}(z).$$

These operators are similar to the vertex operators for the "symplectic bosons" (= even fermions) constructed in [**FMS**] (see also [**FF1**]).

Comparing (5.8.5) with (5.8.6) and Theorem 5.8(d), we obtain the following remarkable identity:

$$(5.8.20) \quad \prod_{n=1}^{\infty} \frac{(1 + sq^n)(1 + s^{-1}q^{n-1})(1 - q^n)^2}{(1 - tq^{n-\frac{1}{2}})(1 - t^{-1}q^{n-\frac{1}{2}})(1 + st^{-1}q^{n-\frac{1}{2}})(1 + s^{-1}tq^{n-\frac{1}{2}})}$$

$$= \sum_{m=0}^{\infty} \frac{t^m q^{\frac{1}{2}m}}{1 + st^{-1}q^{m+\frac{1}{2}}} + \sum_{m=-1}^{-\infty} \frac{s^{-1}t^{m+1}q^{-\frac{1}{2}(m+1)}}{1 + s^{-1}tq^{-m-\frac{1}{2}}}.$$

Substituting in (5.8.20) $s = -uvq^{-1}$, $t = uq^{-\frac{1}{2}}$ brings this identity to a more symmetric form:

$$(5.8.21) \qquad \prod_{n=1}^{\infty} \frac{(1 - uvq^{n-1})(1 - u^{-1}v^{-1}q^n)(1 - q^n)^2}{(1 - uq^{n-1})(1 - u^{-1}q^n)(1 - vq^{n-1})(1 - v^{-1}q^n)}$$

$$= \sum_{m=0}^{\infty} \frac{u^m}{1 - vq^m} - v^{-1} \sum_{m=-1}^{-\infty} \frac{u^m}{1 - v^{-1}q^{-m}} \,.$$

We may expand the right-hand side in the following nice form:

$$(5.8.22) \qquad \left(\sum_{m,n \geq 0} - \sum_{m,n < 0} \right) u^m v^n q^{mn} \,.$$

Identity (5.8.21) is the "denominator identity" (see [**K2**]) for the affine subalgebra $s\ell(2,1)\hat{\,}$. (It may also be obtained as a specialization of Ramanujan's summation formula for the bilateral hypergeometric function $_1\Psi_1$, cf. [**KW2**].) This is, in fact, the simplest among the denominator identities for affine superalgebras discovered in [**KW2**]. It is interesting to note a parallel with Section 5.1. The boson-fermion correspondence based on the Lie algebra $gl(1)\hat{\,}$ lead there to the denominator identity for $s\ell(2)\hat{\,}$, whereas the super boson-fermion correspondence based on the Lie superalgebra $gl(1,1)\hat{\,}$ produced the denominator identity for $s\ell(2,1)\hat{\,}$.

Replacing in 5.8.21 and 5.8.22 q by q^2, u by qz^{-1} and v by z, we obtain the denominator identity for the $N = 2$ superconformal algebra (discussed in the next section) and of $gl(1,1)\hat{\,}$:

$$\prod_{n=1}^{\infty} \frac{(1 - q^n)^2}{(1 - q^{n-1}z)(1 - q^n z^{-1})} = \sum_{n \in \mathbb{Z}} (-1)^n \frac{q^{n(n+1)/2}}{1 - q^n z} \,.$$

I will explain now how to use identity (5.8.21) to obtain some classical results on sums of squares. Let

$$\Box(q)^d = \left(\sum_{j \in \mathbb{Z}} q^{j^2} \right)^d = \sum_{k \in \mathbb{Z}_+} \Box_{d,k} q^k \,.$$

Then, obviously, $\Box_{d,k}$ is the number of representations of k as a sum of d squares of integers (taking into account the order of summands). By Gauss formula (5.1.17) we have:

$$(5.8.23) \qquad \Box(-q)^d = \prod_{j=1}^{\infty} \left(\frac{1 - q^j}{1 + q^j} \right)^d \,.$$

Multiplying both sides of (5.8.21) with the right-hand side (5.8.22) by $(1 - uv)^{-1}(1 - u)(1 - v)$, we obtain:

$$(5.8.24) \quad \prod_{n=1}^{\infty} \frac{(1 - q^n)^2(1 - uvq^n)(1 - u^{-1}v^{-1}q^n)}{(1 - uq^n)(1 - u^{-1}q^n)(1 - vq^n)(1 - v^{-1}q^n)}$$

$$= 1 + \frac{(1 - u)(1 - v)}{1 - uv} \sum_{m,n=1}^{\infty} q^{mn}(u^m v^n - u^{-m}v^{-n}).$$

Letting in this formula $u = v = i$ and replacing q^2 by q, we obtain a formula which goes back to Gauss and Jacobi:

$$(5.8.25) \quad \Box(q)^2 = 1 + 4 \sum_{j,k=1}^{\infty} (-1)^k q^{j(2k-1)}.$$

This formula means that $\Box_{2,k}$ equals the difference between the number of divisors of k congruent to $1 \mod 4$ and the number of divisors of k congruent to $-1 \mod 4$.

Letting in (5.8.24) $u = v = -z$, we obtain:

$$(5.8.26) \quad \prod_{n=1}^{\infty} \frac{(1 - q^n)^2(1 - z^2 q^n)(1 - z^{-2}q^n)}{(1 + zq^n)^2(1 + z^{-1}q^n)^2}$$

$$= 1 + (1 + z) \sum_{m,n=1}^{\infty} (-1)^{m+n} q^{mn} \frac{z^{m+n} - z^{-m-n}}{1 - z}.$$

Letting z tend to 1, we get:

$$\Box(-q)^4 = 1 - 4 \sum_{m,n=1}^{\infty} (-1)^{m+n}(m + n)q^{mn}$$

$$= 1 + 8 \sum_{m,n=1}^{\infty} (-1)^{m+n-1} m q^{mn} = 1 + 8 \sum_{m=1}^{\infty} \frac{m(-q)^m}{1 + q^m}.$$

Changing the sign of q and replacing in the last sum the summation over even integers $2m$ by $\frac{2mq^{2m}}{1-q^{2m}} - \frac{4mq^{4m}}{1-q^{4m}}$, we obtain another of Jacobi's formulas:

$$(5.8.27) \quad \Box(q)^4 = 1 + 8 \sum_{\substack{k=1 \\ 4 \nmid k}}^{\infty} \frac{kq^k}{1 - q^k} = 1 + 8 \sum_{\substack{j,k=1 \\ 4 \nmid k}}^{\infty} kq^{jk},$$

which means that $\Box_{4,k}$ for $k \geq 1$ equals 8 times the sum of divisors of k not divisible by 4.

5.9. Superconformal vertex algebras

The classification of all simple vertex algebras is certainly a hopeless problem as it includes the classification of all non-degenerate integral lattices. Some people impose the condition that a simple vertex algebra V is graded by $\frac{1}{2}\mathbb{Z}_+$:

$$(5.9.1) \qquad V = \oplus_{j \in \frac{1}{2}\mathbb{Z}_+} V^{(j)}, \quad \text{where } V^{(0)} = \mathbb{C}|0\rangle.$$

This excludes lattice vertex algebras associated to indefinite lattices, which is very unfortunate since the vertex algebras associated to Lorentzian lattices provide some of the most spectacular applications of vertex algebras found by Borcherds [**B2**], [**B3**]. But even this restriction includes lattice vertex algebras associated to positive definite integral lattices, and the classification of the latter is still a hopeless problem.

Let V be a simple graded vertex algebra with a gradation (5.9.1). If V is strongly generated by its fields of conformal weight $\frac{1}{2}$, then, due to Theorem 3.6, V is isomorphic to the simple free fermionic vertex algebra $F^k(A)$ (see Section 4.7). Moreover, the same Theorem 3.6 implies that if a vertex algebra V contains $F^k(A)$ as a subalgebra, then $V \simeq F^k(A) \otimes C_V(F^k(A))$. Furthermore, the fields of conformal weight 1 generate an affine vertex subalgebra (cf. Sections 2.6 and 4.7). The simplest result concerning the next case, that of conformal weight $\frac{3}{2}$, is the following (cf. Example 4.10).

LEMMA 5.9. *Let V be a simple conformal vertex algebra strongly generated by a non free odd field of conformal weight $\frac{3}{2}$. Then V is isomorphic to the simple conformal vertex algebra $V^c(NS)$ where NS is a Lie superalgebra spanned by mutually local formal distributions $G(z)$, $L(z) = Y(\nu, z)$ and central element C satisfying the following OPE:*

$$(5.9.2a) \qquad G(z)G(w) \sim \frac{\frac{2}{3}C}{(z-w)^3} + \frac{2L(w)}{z-w},$$

$$(5.9.2b) \qquad L(z)G(w) \sim \frac{\frac{3}{2}G(w)}{(z-w)^2} + \frac{\partial G(w)}{z-w},$$

$$(5.9.2c) \qquad L(z)L(w) \sim \frac{\frac{1}{2}C}{(z-w)^4} + \frac{2L(w)}{(z-w)^2} + \frac{\partial L(w)}{z-w}.$$

PROOF. By the assumptions of the proposition we have

$$V^{(0)} = \mathbb{C}|0\rangle, \quad V^{(1/2)} = V^{(1)} = 0, \quad V^{(3/2)} = \mathbb{C}\tau, \quad V^{(2)} = \mathbb{C}\nu.$$

Hence $L_n \tau = \frac{3}{2}\delta_{n,0}\tau$ for $n \geq 0$, i.e., $G(z) := Y(\tau, z)$ is a primary field of conformal dimension $\frac{3}{2}$, which is equivalent to (5.9.2b). The most general possibility for $G(z)G(w)$ is:

$$G(z)G(w) \sim \frac{2\alpha}{(z-w)^3} + \frac{a(w)}{(z-w)^2} + \frac{\beta L(w)}{z-w}, \quad \alpha, \beta \in \mathbb{C}.$$

In the same way as in the proof of Theorem 2.6a, we show, using locality, that $a(w) = 0$. Since $G(z)$ is not a free field, $\beta \neq 0$ and we may rescale τ to make $\beta = 2$.

It remains to show that $\alpha = \frac{1}{3}$. Let $G(z) = \sum_{n \in \frac{1}{2}+\mathbb{Z}} G_n z^{-n-3/2}$. Then the OPE (5.9.2a–c) are equivalent to the commutation relations (by (2.6.2a)):

$$(5.9.3a) \qquad [G_m, G_n] = 2L_{m+n} + \frac{1}{3}\left(m^2 - \frac{1}{4}\right)\delta_{m,-n}C,$$

$$(5.9.3b) \qquad [G_m, L_n] = \left(m - \frac{n}{2}\right)G_{m+n},$$

$$(5.9.3c) \qquad [L_m, L_n] = (m-n)L_{m+n} + \frac{m^3 - m}{12}\delta_{m,-n}C.$$

We have established all these relations except for identifying the coefficient $\frac{C}{3}$ in (5.9.3a) with α. This follows from the Jacobi identity for the elements G_m, G_n, L_{-m-n}. $\qquad \square$

The Lie superalgebra defined by commutation relations (5.9.3a–c) (with C being a central element) is called the *Neveu-Schwarz algebra* (*NS*) or the $N = 1$ *superconformal Lie algebra* [**NS**].

DEFINITION 5.9. An odd vector τ of a vertex algebra V is called a $N = 1$ *superconformal vector* if the field $G(z) = Y(\tau, z)$ satisfies (5.9.2a and b) with $L(z) = \sum_{n \in \mathbb{Z}} L_n z^{-n-2}$ being a Virasoro field such that $L_{-1}(= G^2_{-\frac{1}{2}}) = T$ and L_0 is diagonalizable.

The following proposition is proved in the same way as Theorem 4.10.

PROPOSITION 5.9. *An odd vector τ of a vertex algebra V is a $N = 1$ supconformal vector iff for $Y(\tau, z) = \sum_{n \in \frac{1}{2}+\mathbb{Z}} G_n z^{-n-\frac{3}{2}}$ the following properties hold:*

(i) $G_{-\frac{1}{2}}\tau = 2\nu$, *where* $Y(\nu, z) = \sum_n L_n z^{-n-2}$ *is such that* $L_{-1} = T$, $L_0\tau = \frac{3}{2}\tau$, L_0 *is diagonalizable,*

(ii) $G_{\frac{3}{2}}\tau = \frac{2}{3}c|0\rangle$ *for some* $c \in \mathbb{C}$,

(iii) $G_k\tau = 0$ *for* $k > \frac{3}{2}$.

In this case ν is a conformal vector. $\qquad\qquad\square$

A vertex algebra V is called $N = 1$ superconformal if it is endowed with a superconformal vector. (Then V is a conformal vertex algebra.)

A large class of $N = 1$ superconformal vertex algebras is provided by the superaffine vertex algebras $V^k(\widehat{\mathfrak{g}}_{\text{super}})$. Namely, the following theorem can be proved in the same way as Theorem 5.7, or simply by making use of the non-commutative Wick formula (3.3.12). Therefore we omit its proof.

THEOREM 5.9. *Let $\mathfrak{g}$ be a simple or commutative finite-dimensional Lie super-algebra with a non-degenerate invariant supersymmetric bilinear form $(.|.)$, and let $\{a^i\}$ and $\{b^i\}$ be dual bases of $\mathfrak{g}$. Then, provided that $k \neq -h^\vee$, the vector*

$$\tau = \frac{1}{k+h^\vee} \left(\sum_i a^i_{-1} b^i_{-\frac{1}{2}} + \frac{2}{3} \sum_{i,j,r} ([a^i, a^j]|a^r) \, b^i_{-\frac{1}{2}} b^j_{-\frac{1}{2}} b^r_{-\frac{1}{2}} \right) |0\rangle$$

is a $N = 1$ superconformal vector of the vertex algebra $V^k(\widehat{\mathfrak{g}}_{\text{super}})$ (and $\widetilde{V}^k(\widehat{\mathfrak{g}}_{\text{super}})$), the corresponding conformal vector being

$$\nu = \frac{1}{k+h^\vee} \left(\frac{1}{2} \sum_i \left(a^i_{-1} b^i_{-1} + a^i_{-3/2} b^i_{-1/2} \right) + \frac{1}{3} \sum_{i,j,r} ([a^i, a^j]|a^r) \, b^i_{-\frac{1}{2}} b^j_{-\frac{1}{2}} b^r_{-1} \right) |0\rangle$$

with central charge

$$\widetilde{c}_k = c_k + \frac{1}{2} \operatorname{sdim} \mathfrak{g} \,. \qquad\qquad\square$$

This construction is usually referred to in physics literature as the *Kac-Todorov model* [**KT1**].

EXAMPLE 5.9a. Let V be the vertex algebra generated by a free bosonic field $\alpha(z) = \sum_{n \in \mathbb{Z}} \alpha_n z^{-n-1}$ and a free (odd) fermionic field $\varphi(z) = \sum_{n \in \frac{1}{2}+\mathbb{Z}} \varphi_n z^{-n-\frac{1}{2}}$ which commute:

$$\alpha(z)\alpha(w) \sim \frac{1}{(z-w)^2}\,, \qquad \varphi(z)\varphi(w) \sim \frac{1}{z-w}\,, \qquad \alpha(z)\varphi(w) \sim 0\,.$$

Then V is a (simple) vertex algebra with a family of $N = 1$ superconformal vectors

$$\tau = \left(\alpha_{-1}\varphi_{-\frac{1}{2}} + \lambda\varphi_{-\frac{3}{2}} \right) |0\rangle\,, \qquad \lambda \in \mathbb{C}\,,$$

the corresponding conformal vector being

$$\nu = \frac{1}{2} \left(\alpha^2_{-1} + \varphi_{-\frac{3}{2}}\varphi_{-\frac{1}{2}} + \lambda\alpha_{-2} \right) |0\rangle$$

with central charge $\frac{3}{2} - 3\lambda^2$. This is proved by a direct calculation using Wick's theorem. The case $\lambda = 0$ (which is Theorem 5.9 for the 1-dimensional Lie algebra $\mathfrak{g}$) goes back to Neveu and Schwarz [**NS**].

EXAMPLE 5.9b. The Neveu-Schwarz algebra NS (defined by (5.9.3a–c)) is spanned by formal distributions $L(z) = \sum_{n \in \mathbb{Z}} L_n z^{-n-2}$, $G(z) = \sum_{n \in \frac{1}{2} + \mathbb{Z}} G_n z^{-n-3/2}$ and C. The derivation $T = \mathrm{ad}L_{-1}$ satisfies (4.7.1), and $H = \mathrm{ad}L_0$ is a Hamiltonian with respect to which $L(z)$, $G(z)$ and C have conformal weights 2, 3/2, and 0 respectively. We have:

$$NS_{--} = \mathbb{C}C + \sum_{m,n \geq -1} (\mathbb{C}L_m + \mathbb{C}G_n), \qquad T(NS_{--}) = \sum_{m,n \geq -1} (\mathbb{C}L_m + \mathbb{C}G_n).$$

Hence (due to Theorem 4.7) the associated universal vertex algebras $\widetilde{V}^c(NS)$ are parameterized by a complex number c ($= \lambda(C)$). These vertex algebras (and their quotients) are superconformal with the $N = 1$ superconformal vector $\tau = G_{-3/2}|0\rangle$ and conformal vector $\nu = L_{-2}|0\rangle$ with central charge c. The vertex algebra $\widetilde{V}^c(NS)$ has a unique simple quotient $V^c(NS)$.

The Neveu-Schwarz algebra is the simplest among the superconformal Lie algebras. Further examples were constructed in [**A**], [**K1**], [**KL2**], [**CK1**]. We define a *superconformal Lie algebra* as a Lie superalgebra $\mathfrak{g}$ for which the following three properties hold:

(a) $\mathfrak{g}$ is spanned by a finite family of pairwise local formal distributions $\{a^\alpha(z)\}$;

(b) $\mathfrak{g}$ is simple in the sense that no submodule of the $\mathbb{C}[\partial]$-module $\sum_\alpha \mathbb{C}[\partial]a^\alpha(z)$ spans a nontrivial ideal of $\mathfrak{g}$;

(c) one of the members of the family is the Virasoro formal distribution $L(z)$ satisfying the properties:

$$[L_{-1}, a^\alpha(z)] = \partial a^\alpha(z), \qquad [L_0, a^\alpha(z)] = (z\partial + \Delta_\alpha)a^\alpha(z), \qquad \Delta_\alpha \in \mathbb{C}.$$

It follows easily from a very difficult theorem of Mathieu [**M**] that the only simple graded Lie algebras spanned by a finite number of pairwise local formal distributions are the Virasoro algebra and the affine Kac-Moody algebras modulo their centers. It seems plausible that the condition that the Lie algebra should be graded is superfluous, i.e. any Lie algebra satisfying condition (a) and (b) is either Virasoro or an affine Kac-Moody algebra modulo the center; cf. Conjecture 2.7. A conjectural list of all superconformal Lie algebras was given in [**KL2**]. However, it has been discovered recently that one should add to this list a new superconformal Lie algebra, denoted by CK_6, which is spanned by 16 even and 16 odd pairwise local formal distributions [**CK1**].

The simplest after the Neveu-Schwarz algebra is the celebrated $N = 2$ superconformal Lie algebra. It is a graded superalgebra spanned by a central element C, a Virasoro formal distribution $L(z)$, an even formal distribution $J(z)$ primary with respect to $L(z)$ of conformal weight 1, and two odd primary with respect to $L(z)$ formal distributions $G^+(z)$ and $G^-(z)$ of conformal weight $3/2$. The remaining OPE are as follows:

$$(5.9.4\text{a}) \quad J(z)J(w) \sim \frac{C/3}{(z-w)^2}, \quad G^\pm(z)G^\pm(w) \sim 0, \quad J(z)G^\pm(w) \sim \pm\frac{G^\pm(w)}{z-w},$$

$$(5.9.4\text{b}) \quad G^+(z)G^-(w) \sim \frac{C/3}{(z-w)^3} + \frac{J(w)}{(z-w)^2} + \frac{L(w) + \frac{1}{2}\partial J(w)}{z-w}.$$

We denote this superalgebra by $N2$. Note that the superalgebra $N2$ contains the Neveu-Schwarz subalgebra spanned by C, $L(z)$ and $G(z) = G^-(z) + G^+(z)$, and another one spanned by C, $L(z)$ and $\bar{G}(z) = i(G^+(z) - G^-(z))$.

A vertex algebra is called $N = 2$ superconformal if it has two odd vectors $\tau^\pm$ such that the fields $G^\pm(z) = Y(\tau^\pm, z)$ satisfy the OPE of $N = 2$ superconformal algebra with $L_{-1} \left(= [G^+_{-\frac{1}{2}}, G^-_{-\frac{1}{2}}]\right) = T$, and J_0 and L_0 are diagonizable.

Of course one constructs the $N = 2$ superconformal vertex algebras $V^c(N2)$ in the same way as in Example 5.9b. A more explicit example is the following.

EXAMPLE 5.9c. Let $\mathbb{Z}(3)$ be the lattice of rank 1 with the bilinear form $(m|n) = 3mn$. Then the simple lattice vertex algebra $V_{\mathbb{Z}(3)}$ is isomorphic to the $N = 2$ superconformal vertex algebra $V^1(N2)$. Indeed, the fields $G^\pm(z) = \frac{1}{\sqrt{3}}\Gamma_\pm(z)$, $J(z) = \frac{1}{3}1(z)$, $L(z) = \frac{1}{6} : 1(z)^2 :$ strongly generate the vertex algebra $V_{\mathbb{Z}(3)}$ and obey the OPE of $N2$ with $c = 1$. This follows from the general OPE formulas (5.5.13), (5.5.14) and (5.5.18), and Proposition 5.5.

EXAMPLE 5.9d. (cf. [**OS**]) Let V be the vertex algebra generated by a pair of free charged bosonic fields $\alpha^\pm(z) = \sum_{n \in \mathbb{Z}} \alpha^\pm_n z^{-n-1}$ and a pair of free charged (odd) fermionic fields $\psi^\pm(z) = \sum_{n \in \frac{1}{2}+\mathbb{Z}} \psi^\pm_n z^{-n-1/2}$ such that:

$$\alpha^\pm(z)\alpha^\mp(w) \sim \frac{1}{(z-w)^2}, \quad \psi^\pm(z)\psi^\mp(w) \sim \frac{1}{z-w},$$

OPE for all other pairs of fields ~ 0. Then V is a (simple) vertex algebra with a family of $N = 2$ superconformal vectors

$$\tau^{\pm} = \left(\alpha^{\pm}_{-1} \psi^{\pm}_{-\frac{1}{2}} \pm \lambda \psi^{\pm}_{-\frac{3}{2}} \right) |0\rangle \,,$$

the vector corresponding to the current $J(z)$ being

$$j = \left(\psi^{+}_{-\frac{1}{2}} \psi^{-}_{-\frac{1}{2}} - \lambda \alpha^{+}_{-1} - \lambda \alpha^{-}_{-1} \right) |0\rangle \,,$$

and the corresponding conformal vector being

$$\nu = \left(\alpha^{+}_{-1} \alpha^{-}_{-1} + \frac{1}{2} \psi^{+}_{-3/2} \psi^{-}_{-1/2} + \frac{1}{2} \psi^{-}_{-3/2} \psi^{+}_{-1/2} - \frac{\lambda}{2} \alpha^{+}_{-2} + \frac{\lambda}{2} \alpha^{-}_{-2} \right) |0\rangle \,,$$

with central charge $3 + 6\lambda^2$.

Kazama and Suzuki [**KS**] have found necessary and sufficient condition for a coset model of a $N = 1$ superconformal vertex algebra $V^c(\widehat{\mathfrak{g}}_{\mathrm{super}})$ to admit a $N = 2$ superconformal vector.

Another point of view at $N = 1$ superconformal vertex algebra V is as follows. Introduce *superfields*:

$$(5.9.5) \qquad \overset{s}{Y}(a, z, \xi) = Y(a, z) + \xi Y(G_{-\frac{1}{2}} a, z) \,,$$

where ξ is an odd indeterminate, $\xi^2 = 0$. Using that, by (4.9.9),

$$(5.9.6) \qquad [G_{-\frac{1}{2}}, Y(a, z)] = Y(G_{-\frac{1}{2}} a, z) \,,$$

we obtain

$$(5.9.7) \qquad [G_{-\frac{1}{2}}, \overset{s}{Y}(a, z, \xi)] = (\partial_{\xi} + \xi \partial_z) \overset{s}{Y}(a, z, \xi) \,.$$

This leads us to a $N = 1$ *vertex superalgebra* V (generalizing that of a $N = 1$ superconformal vertex algebra) defined by a vector $|0\rangle \in V$, odd operator $G \; (= G_{-\frac{1}{2}})$ on V, and superfields $\overset{s}{Y}(a, z, \xi)$ $(a \in V)$ such that the "odd translation covariance" axiom (5.9.7) holds, the usual locality axiom holds and the obvious modification of vacuum axioms hold:

$$(5.9.8) \qquad G|0\rangle = 0 \,, \qquad \overset{s}{Y}(|0\rangle, z, \xi) = I_V \,, \qquad \overset{s}{Y}(a, z, \xi)|0\rangle|_{z=0, \xi=0} = a \,.$$

Most of the formulas and results for vertex algebras remain valid for $N = 1$ vertex superalgebras if one replaces $Y(a, z)$ by $\overset{s}{Y}(a, z, \xi)$, including (4.4.5–4.4.7) and Borcherds OPE formula (and identity). For example, the OPE formula reads:

$$\overset{s}{Y}(a, z, \xi)\overset{s}{Y}(b, w, \xi) \sim \sum_{n \in \mathbb{Z}_+} \frac{\overset{s}{Y}(a_{(n)}b, w, \xi)}{(z - w)^{n+1}} \, .$$

It is also easy to see that $N = 1$ vertex superalgebra is precisely a vertex algebra with an odd operator $G_{-\frac{1}{2}}$ such that $G^2_{-\frac{1}{2}} = T$ and (5.9.6) holds. (Then superfields are defined by (5.9.5).) This latter definition generalizes to an arbitrary $N = n$. A $N = n$ *vertex superalgebra* is a vertex algebra with n odd operators $G^{(i)}$ satisfying for all $i, j = 1, \dots, n$ the following two conditions:

$$(5.9.9) \qquad\qquad [G^{(i)}, Y(a, z)] = Y(G^{(i)}a, z) \, ,$$

$$(5.9.10) \qquad\qquad [G^{(i)}, G^{(j)}] = 2\delta_{ij}T \, .$$

The $N = 2$ superconformal vertex algebra is $N = 2$ vertex superalgebra with $G^{(1)} = G^+_{-\frac{1}{2}} + G^-_{-\frac{1}{2}}$, $G^{(2)} = i(G^+_{-\frac{1}{2}} - G^-_{-\frac{1}{2}})$.

EXAMPLE 5.9e. $V^k(\widehat{\mathfrak{g}}_{\text{super}})$ is a $N = 1$ vertex superalgebra (for all k) generated by the fields

$$\bar{a}(z, \xi) = \sum_n (at^n\theta)z^{-n-1} + \sum_n (at^n)z^{-n-1}\xi, \quad a \in \mathfrak{g} \, .$$

The operator G is induced by the derivation $\partial_\theta - \theta\partial_t$ of $\widehat{\mathfrak{g}}_{\text{super}}$ (since $(\partial_\theta - \theta\partial_t)\bar{a}(z, \xi) = (\partial_\xi + \xi\partial_z)\bar{a}(z, \xi)$). This operator coincides with $G_{-\frac{1}{2}}$ given by Theorem 5.9 for $k \neq -h^\vee$.

For a $N = n$ vertex superalgebra the superfields are constructed in the same way as for $N = 1$ (where the ξ_i are anticommuting indeterminates):

$$\overset{s}{Y}(a, z, \xi_1, \dots, \xi_n) = \sum_{\substack{0 \leq r \leq n \\ 1 \leq i_1 < \cdots < i_r \leq n}} Y(G^{(i_1)} \cdots G^{(i_r)}a, z)\xi_{i_1} \cdots \xi_{i_r} \, .$$

Then (5.9.7) holds for each ξ_i (= the "odd translation covariance" axiom) and the usual locality axiom and the obvious modification of the vacuum axioms (5.9.8) hold as well. It is easy to see that these axioms give an equivalent definition of a $N = n$ vertex superalgebra. Example 5.9e generalizes to an arbitrary n in an obvious way.

REMARK 5.9. Condition (5.9.10) puts quite stringent constraints on the number of generating fields. For example, let $\mathfrak{g}$ be a superconformal Lie algebra spanned by the coefficients of a finitely generated $\mathbb{C}[\partial]$-module with a basis consisting of pairwise local formal distributions $a^\alpha(z)$. Suppose that $a^\beta(z)_{(0)}a^\gamma(z) = L(z) + \partial\varphi(z)$ for some indices β, γ and some formal distribution $\varphi(z)$ (cf. (5.9.10)). Then one has:

$$(5.9.11) \qquad\qquad \#(\text{even } a^\alpha) = \#(\text{odd } a^\alpha).$$

To prove this relation, consider the $\mathbb{C}[\partial]$-module $\mathcal{A}$ spanned by the $a^\alpha(z)$. Then $A := \mathcal{A}/\partial\mathcal{A}$ is a Lie superalgebra with respect to the 0-th product and $\mathcal{A}$ is a left module over A defined by this product (see Section 2.3) such that:

$$(5.9.12) \qquad\qquad L \cdot a^\alpha = \partial a^\alpha.$$

But $L = a^\beta_{(0)}a^\gamma$ in A, hence its supertrace $\operatorname{str} L$ must be zero. On the other hand, $\operatorname{str} L = (\#(\text{even } a^\alpha) - \#(\text{odd } a^\alpha))\,\partial$ by (5.9.12). A special case of (5.9.11) was obtained in [**RS**] as a result of a lengthy calculation.

5.10. On classification of conformal superalgebras

As we have seen in the previous section, the superconformal Lie algebras give rise to some of the most important vertex algebras. On the other hand, according to Section 2.7, the classification of finite formal distribution Lie superalgebras reduces to the classification of finite conformal superalgebras. Here we shall discuss briefly the latter problem.

First, we recall the two examples that arose in the previous sections (disregarding the central terms).

The simplest is the Neveu-Schwarz (or $N = 1$) conformal superalgebra $R = \mathbb{C}[\partial]L + \mathbb{C}[\partial]G$, where the generator L is even, the generator G is odd, and all non-trivial n-th products between them are as follows:

$$(5.10.1a) \quad L_{(0)}L = \partial L,\; L_{(1)}L = 2L,$$

$$(5.10.1b) \quad L_{(0)}G = \partial G,\; G_{(0)}L = \frac{1}{2}\partial G,\; L_{(1)}G = G_{(1)}L = \frac{3}{2}G,\; G_{(0)}G = 2L.$$

Formula (5.10.1a) shows that the even subalgebra $\mathbb{C}[\partial]L$ is nothing else but the Virasoro conformal algebra. The central extension of the Neveu-Schwarz conformal superalgebra (cf. (5.10.3)) is given by

$$(5.10.2) \qquad \alpha_2(G,G) = \frac{2}{3}c, \quad \alpha_3(L,L) = \frac{1}{2}c.$$

Here and further we are writing only non-zero values of the cocycle on generators. It is easy to snow that any 2-cocycle is equivalent to (5.10.2).

The following definition (which is a counterpart of Definition 2.6a, see also Section 4.10) facilitates the description and classification of conformal superalgebras.

DEFINITION 5.10. Let R be a conformal superalgebra and let L be an even element of R. An element $a \in R$ is called an *eigenvector* with respect to L of *conformal weight* $\Delta \in \mathbb{C}$ if

$$(5.10.3) \qquad L_{(0)}a = \partial a, \quad L_{(1)}a = \Delta a.$$

An eigenvector a with respect to L is called *primary* if $L_{(m)}a = 0$ for $m > 1$. The conformal superalgebra R is called *graded* by L if it has a basis over $\mathbb{C}$ consisting of eigenvectors with respect to L.

Note that, by (C2), we have for a primary eigenvector a of conformal weight Δ with respect to L:

$$(5.10.4) \qquad a_{(0)}L = (\Delta - 1)\partial a, \quad a_{(1)}L = \Delta a, \quad a_{(m)}L = 0 \text{ for } m \geq 1.$$

Note also the following simple properties of the conformal weight (cf. Corollary 2.6).

LEMMA 5.10. *If elements a and a' are eigenvectors with respect to L of conformal weight Δ and Δ', then:*

(a) *∂a is an eigenvector of conformal weight $\Delta + 1$,*

(b) *$a_{(n)}a'$ is an eigenvector of conformal weight $\Delta + \Delta' - n - 1$.*

PROOF. (a) follows from (C1) and (2.7.3). (b) follows from (C3) for $m = 0$ and $m = 1$, $n \in \mathbb{Z}_+$ and $a = L$, $b = a$, $c = a'$, and from (C1) . $\qquad \square$

The second example is the $N = 2$ conformal superalgebra $R = R_{\bar{0}} \oplus R_{\bar{1}}$ where both $R_{\bar{0}}$ and $R_{\bar{1}}$ are free $\mathbb{C}[\partial]$-modules of rank 2:

$$R_{\bar{0}} = \mathbb{C}[\partial]L \oplus \mathbb{C}[\partial]J, \quad R_{\bar{1}} = \mathbb{C}[\partial]G^+ \oplus \mathbb{C}[\partial]G^- .$$

The element L is a Virasoro element (i.e.(2.7.18)) holds), elements J and $G^\pm$ are primary eigenvectors with respect to L of conformal weight 1 and $3/2$, and all the remaining non-trivial products (up to changing the order) are as follows:

$$(5.10.5) \qquad J_{(0)}G^\pm = \pm G^\pm, \quad G^+_{(0)}G^- = L + \frac{1}{2}\partial J, \quad G^+_{(1)}G^- = J.$$

The $N = 2$ conformal superalgebra has a unique (up to equivalence) 2-cocycle given by

$$(5.10.6) \qquad \alpha_1(J,J) = \frac{1}{3}c, \quad \alpha_2(G^+, G^-) = \frac{1}{3}c, \quad \alpha_3(L,L) = \frac{1}{2}c.$$

The next most important example is the following $N = 4$ conformal superalgebra. It is convenient to use for its description the basis of $sl_2(\mathbb{C})$ consisting of Pauli matrices $\sigma^s = (\sigma^s_{ab})_{a,b\in\{1,2\}}$, $s = 1, 2, 3$:

$$\sigma^1 = \begin{pmatrix} 0 & 1 \\ 1 & 0 \end{pmatrix}, \quad \sigma^2 = \begin{pmatrix} 0 & -i \\ i & 0 \end{pmatrix}, \quad \sigma^3 = \begin{pmatrix} 1 & 0 \\ 0 & -1 \end{pmatrix}.$$

The even part of the $N = 4$ conformal superalgebra is $R_{\bar{0}} = \mathbb{C}[\partial]L \oplus \left(\bigoplus_{s=1}^3 \mathbb{C}[\partial]J^s\right)$, where L is a Virasoro element, the J^s are primary eigenvectors with respect to L of conformal weight 1 and $\bigoplus_{s=1}^3 \mathbb{C}[\partial]J^s$ is the current conformal algebra associated to $sl_2(\mathbb{C})$ with the basis $J^s = \frac{1}{2}\sigma^s$. The odd part too is a free $\mathbb{C}[\partial]$-module of rank 4: $R_{\bar{1}} = \mathbb{C}[\partial]G^1 + \mathbb{C}[\partial]G^2 + \mathbb{C}[\partial]\bar{G}^1 + \mathbb{C}[\partial]\bar{G}^2$, all four elements G^1, G^2, $\bar{G}^1$, and $\bar{G}^2$ being primary eigenvectors with respect to L of conformal weight $3/2$. The remaining (up to the order) non-trivial n-th products are as follows:

$$(5.10.7a) \qquad J^s_{(0)}G^a = \frac{1}{2}\sum_b \sigma^s_{ab}G^b, \quad J^s_{(0)}\bar{G}^a = -\frac{1}{2}\sum_b \sigma^s_{b,a}\bar{G}^b,$$

$$(5.10.7b) \qquad G^a_{(0)}\bar{G}^b = 2\delta_{ab}L + 2\sum_s \sigma^s_{ab}\partial J^s, \quad G^a_{(1)}\bar{G}^b = 4\sum_s \sigma^s_{ab}J^s.$$

The unique, up to equivalence, 2-cocycle is given by

$$(5.10.8) \qquad \alpha_1(J^s, J^t) = \frac{c}{6}\mathrm{tr}J^s J^t, \quad \alpha_2(G^a, \bar{G}^b) = \frac{2}{3}\delta_{ab}c, \quad \alpha_3(L,L) = \frac{c}{2}.$$

Now we can state the main theorem of this section.

THEOREM 5.10. *Let R be a graded by an element L simple finite conformal superalgebra. Suppose that in addition the following conditions hold:*

(i) *L is a Virasoro element,*

(ii) *eigenvectors of conformal weight* 1 *along with* L *generate the* $\mathbb{C}[\partial]$-*module*
   $R_{\bar{0}}$,

(iii) *eigenvectors of conformal weight* 3/2 *generate the* $\mathbb{C}[\partial]$-*module* $R_{\bar{1}}$.

Then R *is isomorphic to one of the following four conformal superalgebras: Virasoro, Neveu-Schwarz,* $N = 2$, *and* $N = 4$.

PROOF. Let $\mathfrak{g} \subset R_{\bar{0}}$ (resp. $V \subset R_{\bar{1}}$) denote the subspace over $\mathbb{C}$ of all primary eigenvectors of conformal weight 1 (resp. 3/2). Then, due to (ii), (iii), and Corollary 2.7, we have:

$$R_{\bar{0}} = \mathbb{C}[\partial]L \oplus (\mathbb{C}[\partial] \otimes_{\mathbb{C}} \mathfrak{g}), \quad R_{\bar{1}} = \mathbb{C}[\partial] \otimes_{\mathbb{C}} V.$$

Due to Lemma 5.10, $\mathfrak{g}$ is a Lie algebra with respect to 0-th product and we have a representation π of $\mathfrak{g}$ on V defined by

$$\pi(g)v = g_{(0)}v, \quad g \in \mathfrak{g}, \quad v \in V.$$

The remaining non-trivial products, due to Lemma 5.10, have the following form $(u, v \in V)$:

$$
\begin{aligned}
u_{(0)}v &= 2(u,v)L + \partial\varphi(u \otimes v), \\
u_{(1)}v &= \psi(u \otimes v),
\end{aligned}
$$

where (u,v) is a symmetric $\mathbb{C}$-valued bilinear form on V, invariant with respect to the representation π of $\mathfrak{g}$ on V, and φ and ψ are homomorphisms of $\mathfrak{g}$-modules $V \otimes V \to \mathrm{ad}\,\mathfrak{g}$.

Due to (C3) we have for $u, v \in V$:

$$L_{(2)}(u_{(0)}v) = (L_{(0)}u)_{(2)}v + 2(L_{(1)}u)_{(1)}v = u_{(1)}v.$$

It follows that

$$(5.10.9) \qquad\qquad \psi = 2\varphi.$$

Note that the bilinear form $(\,,\,)$ on V is non-degenerate (otherwise $\mathbb{C}[\partial] \otimes_{\mathbb{C}} (\mathfrak{g} + \mathrm{Ker}(\,,\,))$ is an ideal of R) and that the representation π of $\mathfrak{g}$ on V is faithful (its kernel is an ideal of R).

In the case $R_{\bar{1}} = 0$, the conformal algebra $R = R_{\bar{0}}$ is the Virasoro conformal algebra since $\mathbb{C}[\partial] \otimes_{\mathbb{C}} \mathfrak{g}$ is an ideal of R in this case.

Let now $R_{\bar{1}} \neq 0$ and consider the $R/\partial R$-module R given by 0-th product. Then, due to the non-degeneracy of the bilinear form $(.,.)$ on V, we may apply the argument of Remark 5.10 to get

$$(5.10.10) \qquad\qquad \dim \mathfrak{g} = \dim V - 1 .$$

Let $u \in V$ be such that $(u, u) = 1$, and let $u^{\perp}$ be the orthogonal complement to $\mathbb{C}u$ in V. Then in the basis L, $\mathfrak{g}$, u, $u^{\perp}$ of R the matrix of the element u (viewed as an element of $R/\partial R$ acting on R by 0-th product) looks as follows:

$$u = \left(\begin{array}{cc|cc} 0 & 0 & 2 & 0 \\ 0 & 0 & \alpha\partial & \lambda\partial \\ \hline \partial/2 & \gamma & 0 & 0 \\ 0 & \nu & 0 & 0 \end{array}\right)$$

where λ and ν are $\dim \mathfrak{g} \times \dim \mathfrak{g}$ matrices over $\mathbb{C}$. Since R is a $R/\partial R$-module, we deduce that the square of this matrix is ∂I, where I is the identity matrix (cf.Remark 5.10). It follows that $\alpha = 0$, $\gamma = 0$, $\lambda\nu = I_{\dim \mathfrak{g}}$. In particular, the matrix ν is invertible, which implies that

$$(5.10.11) \qquad\qquad \dim \pi(\mathfrak{g})u = \dim \mathfrak{g} .$$

Let G be the connected simply connected Lie group with the Lie algebra $\mathfrak{g}$. Due to (5.10.10) and (5.10.11) the group G acts transitively on the quadric $(u, u) = 1$ with a discrete stabilizer. If $\dim V = 1$ (resp. $= 2$), then, by (5.10.10), $\dim \mathfrak{g} = 0$ (resp. $= 1$) and it is easy to see, using also (5.10.9), that R is isomorphic to the Neveu-Schwarz (resp. $N = 2$) conformal superalgebra.

Let now $N = \dim V > 2$. Then the quadric $(u, u) = 1$ in V, being homeomorphic to the direct product of the $N - 1$-dimensional sphere and $\mathbb{R}^{N-1}$, is simply connected and hence is homeomorphic to G. But this is possible only for $N = 4$ in which case $\mathfrak{g} \simeq sl_2(\mathbb{C})$. Since π is an orthogonal 4-dimensional representation of $sl_2(\mathbb{C})$ which has a 3-dimensional orbit, the only possibility for π is the direct sum of two 2-dimensional irreducible representations. It is easy to conclude now that in this case R is the $N = 4$ conformal superalgebra. $\qquad\qquad\square$

REMARK 5.10. One can prove by a similar method, using the classification of complex Lie groups acting transitively on quadrics, the following stronger result [**K3**]: Suppose that condition (iii) of Theorem 5.10 is replaced by a weaker condition:

(iii') eigenvectors of conformal weights 3/2 and 1/2 generate the $\mathbb{C}[\partial]$-module $R_{\bar{1}}$.

Then the complete list is obtained by adding to the list of Theorem 5.10 the well-known $N = 3$ conformal superalgebra K_3, of rank 8 over $\mathbb{C}[\partial]$, the less known conformal superalgebra W_2 of rank 12 over $\mathbb{C}[\partial]$ and the new conformal superalgebra CK_6 of rank 32 over $\mathbb{C}[\partial]$ constructed in [**CK1**]; see below. (The algebras K_3 and W_2 admit a unique up to equivalence non-trivial central extension, and the algebra CK_6 admits no non-trivial central extensions.) Under stronger assumptions, which exclude W_2 and CK_6, a similar result was stated in [**RS**], but the proof there is not quite correct.

M. Wakimoto and myself have shown that the only simple conformal superalgebra R with rank $R_{\bar{0}}$ = rank $R_{\bar{1}}$ = 1 is the Neveu-Schwarz algebra. Using methods of [**DK**] and results of [**K4**], I was able to give a complete classification of simple finite conformal superalgebras, proving thereby Conjecture 5.9 from the first edition of this book. The result is stated below.

The list of all finite conformal superalgebras is much richer than that of finite conformal algebras. First, there are many more simple finite-dimensional Lie superalgebras (classified in [**K1**]), and the associated conformal superalgebra is finite and simple. Second, there are many "superizations" of the Virasoro conformal algebra described below. They are associated to superconformal algebras constructed in [**K1**], [**KL2**] and [**CK1**] (cf. [**K3**]).

Let $\Lambda(N)$ denote the Grassmann algebra over $\mathbb{C}$ in N indeterminates $\xi_1, \ldots, \xi_N$. Let $W(N)$ be the Lie superalgebra of all derivations of superalgebra $\Lambda(N)$. It consists of all linear differential operators $\sum_{i=1}^{N} P_i \partial_i$, where $P_i \in \Lambda(N)$ and ∂_i stands for the partial derivative by ξ_i.

The first series of examples is the series conformal superalgebras W_N of rank $(N + 1)2^N$:

$$W_N = \mathbb{C}[\partial] \otimes_{\mathbb{C}} (W(N) \oplus \Lambda(N))$$

with the following products ($a, b \in W(N)$, $f, g \in \Lambda(N)$):

$$a_{(j)}b = \delta_{j0}[a,b], \quad a_{(0)}f = a(f), \quad a_{(j)}f = -\delta_{j1}(-1)^{p(a)p(f)}fa \text{ if } j \geq 1,$$

$$f_{(0)}g = -\partial(fg), \quad f_{(j)}g = -2\delta_{j1}fg \text{ if } j \geq 1.$$

For an element $D = \sum_{i=1}^{N} P_i(\partial, \xi)\partial_i + f(\partial, \xi) \in W_N$ define the $a - twisted$ *divergence* by the formula

$$\mathrm{div}_a D = \sum_{i=1}^{N}(-1)^{p(P_i)}\partial_i P_i - (\partial + a)f, \quad a \in \mathbb{C}$$

The second series of examples is

$$S_{N,a} = \{D \in W_N \mid \mathrm{div}_a D = 0\}.$$

This is a family of simple conformal superalgebras of rank $N2^N$ if $N \geq 2$, $a \in \mathbb{C}$.

The third series of examples is a deformation of $S_N = S_{N,0}$ which exists for each even positive N:

$$\tilde{S}_N = \{D \in W_N | \mathrm{div}(1 + \xi_1 \ldots \xi_N)D = 0\}.$$

The fourth series of examples is K_N. It is also a subalgebra of W_N (of rank 2^N), but it is more convenient to describe it as follows:

$$K_N = \mathbb{C}[\partial] \otimes_{\mathbb{C}} \Lambda(N)$$

with the following products ($f, g \in \Lambda(N)$):

$$f_{(0)}g = \left(\frac{1}{2}|f| - 1\right)\partial fg + \frac{1}{2}(-1)^{|f|}\sum_{i=1}^{N}(\partial_i f)(\partial_i g),$$

$$f_{(j)}g = \left(\frac{1}{2}(|f| + |g|) - 2\right)\delta_{j1}fg \text{ if } j \geq 1.$$

We assume here that f and g are homogeneous elements of degrees $|f|$ and $|g|$ in the gradation defined by $\deg \xi_i = 1$ for all i. The conformal superalgebra K_N is simple for all $N \geq 0$ except for $N = 4$. The derived algebra K_4' is simple and $K_4/K_4' = \mathbb{C}$.

These four series include all well-known examples. Thus, $W_0 \simeq K_0$ is the Virasoro conformal algebra, K_1 is the Neveu-Schwarz conformal superalgebra, $K_2 \simeq W_1$ and K_3 are the $N = 2$ and $N = 3$ conformal superalgebras respectively, S_2 is the $N = 4$ conformal superalgebra. Nevertheless there is one exceptional example constructed in [**CK2**]. It is the subalgebra CK_6 of rank 32 in the conformal super-algebra K_6 spanned over $\mathbb{C}[\partial]$ by the following elements:

$$1 - \alpha\partial^3\nu, \quad \xi_i - \alpha\partial^2\xi_i^*, \quad \xi_i\xi_j - \alpha\partial(\xi_i\xi_j)^*, \quad \xi_i\xi_j\xi_k - \alpha(\xi_i\xi_j\xi_k)^*.$$

Here $\alpha^2 = -1$, $\nu = \xi_1\xi_2\cdots\xi_6$, $(\xi_{i_1}\xi_{i_2}\cdots)^* = \partial_{i_1}\partial_{i_2}\cdots\nu$.

Now we can state the classificational result (cf. [**K4**]): *Any simple finite conformal superalgebra is isomorphic either to the current conformal superalgebra associated to a simple finite-dimensional Lie superalgebra (which are classified in* [**K1**]*), or to one of the following conformal superalgebras* $(N \in \mathbb{Z}_+)$: $W_N, S_{N+2,a}$, $\tilde{S}_{N+2}(N\text{even}), K_N(N \neq 4), K_4', CK_6$.

Bibliography

[AM]   D. Adamovich and A. Milas. Vertex operator algebras associated to modular inveriant representations for $A_1^{(1)}$. *Math. Res. Lett.*, **2** (1995), 563–575.

[A]   M. Ademollo, L. Brink, A. D'Adde, R. D'Auria, E. Napolitano, S. Sciuto, E. Del Guidiche, P. Di Vecchia, S. Ferrara, F. Gliozzi, R. Musto and R. Petturino. Supersymmetric strings and colour confinement. *Phys. Lett.*, **B62** (1976), 105–110.

[BBSS] F. A. Bais, P. Bouwknegt, K.Schoutens, M.Surridge. Extensions of the Virasoro algebra constructed from Kac-Moody algebras using higher order Casimir invariants. *Nucl. Phys.*, **B304** (1988), 348–370.

[BKV]   B. Bakalov, V. G. Kac and A. A. Voronov. Cohomology of conformal algebras, preprint

[BD]   A. A. Beilinson and V. G. Drinfeld. Chiral algebras, preprint

[Be]   M. J. Bergvelt, A note on super Fock space. *J. Math. Phys.*, **30** (1983), 812–815.

[BPZ]   A. Belavin, A. Polyakov, and A. Zamolodchikov. Infinite conformal symmetries in two-dimensional quantum field theory. *Nucl. Phys.*, **B241** (1984), 333–380.

[B1]   R. Borcherds. Vertex algebras, Kac-Moody algebras, and the Monster. *Proc. Natl. Acad. Sci. USA*, **83** (1986), 3068–3071.

[B2]   R. Borcherds. Monstrous moonshine and monstrous Lie superalgebras. *Invent. Math.*, **109** (1992), 405–444.

[B3]   R. Borcherds. Automorphic forms on $O_{s+2,2}(\mathbb{R})$ and infinite products. *Invent. Math.*, **120** (1995), 161–213.

[B4]   R. Borcherds. Vertex algebras, preprint

[Bo]   P. Bouwknegt. Extended conformal algebras from Kac-Moody algebras. In *Infinite-dimensional Lie algebras and groups*, V. G. Kac, ed., *Adv. Ser. in Math. Phys.* **7** (1989), World Scientific, 527–555.

[BS]   P. Bouwknegt, K. Schoutens, eds. W-symmetry. *Adv. Ser. in Math. Phys.*, **22**, World Scientific, 1995.

[CK1]   S.-J. Cheng and V. G. Kac. A new $N = 6$ superconformal algebra. *Commun. Math. Phys.* **186** (1997), 219–231.

[CK2]   S.-J. Cheng and V. G. Kac. Conformal modules, Asian J. of Math., 1 (1997),181–193. Erratum, *Asian J. Math.*, **2** (1998), 153–156.

194   BIBLIOGRAPHY

[CKW]   S.-J. Cheng, V. G. Kac and M. Wakimoto. Extensions of conformal modules. In *Topological field theory, primitive forms and related topics, Proceedings of Taniguchi symposia*, Progress in Math., Birkhäuser, 1998.

[DK]   A. D'Andrea and V. G. Kac. Structure theory of finite conformal algebras, *Selecta Math.* (1998).

[DJKM]   E. Date, M. Jimbo, M. Kashiwara, and T. Miwa. Transformation groups for soliton equations. In *Nonlinear integrable systems—classical theory and quantum theory.* M. Jimbo and T. Miwa, editors, World Scientific, 1983, 39–120.

[DVVV]   R. Dijkgraaf, C. Vafa, E. Verlinde, and H. Verlinde. Operator algebras of orbifold models. *Comm. Math. Phys.*, **123** (1989), 485–526.

[DL]   C. Dong and J. Lepowsky. Generalized vertex algebras and relative vertex operators, *Progress in Math.* **112**, Birkhauser, Boston, 1993.

[DLM]   C. Dong, H. Li and G. Mason. Compact automorphism groups of vertex operator algebras, *Int. Math. Res. Notices* **18** (1996), 913–921.

[DGM]   L. Dolan, P. Goddard and P. Montague. Conformal field theory of twisted vertex operators, *Nucl. Phys.* **B338** (1990), 529–601.

[Fe]   B. L. Feigin, *On the cohomology of the Lie algebra of vector fields and of the current algebra*, Selecta Math. Soviet. **7** (1988), 49–62.

[FeF]   B. L. Feigin and D. B. Fuchs, *Homology of the Lie algebra of vector fields on the line.* (Russian) Funkc. Anal. i Pril. **14** (1980) no. 3, 45–60

[FF1]   B. Feigin and E. Frenkel. Semi-infinite Weil complex and the Virasoro algebra. *Commun. Math. Phys.* **137** (1991), 617–639.

[FF2]   B. Feigin, E. Frenkel. Affine Kac-Moody algebras at the critical level and Gelfand-Dikii algebras, in Infinite Analysis, A. Tsuchiya et all, editors, *Adv. Ser. in Math. Phys.* **16** (1992), World Scientific, 197–215.

[FF3]   B. Feigin, E. Frenkel. Kac-Moody groups and integrability of soliton equations. *Invent. Math.*, **120** (1995), 379–408.

[FKRW]   E. Frenkel, V. Kac, A. Radul, and W. Wang. $W_{1+\infty}$ and $W(gl_N)$ with central charge N. *Comm. Math. Phys.*, **170** (1995), 337–357.

[FK]   I. B. Frenkel and V. G. Kac. Basic representations of affine Lie algebras and dual resonance models. *Invent. Math.*, **62** (1980), 23–66.

[FHL]   I. B. Frenkel, Y. Huang, and J. Lepowsky. On axiomatic approaches to vertex operator algebras and modules. *Mem. Amer. Math. Soc.*, **104** No 494 (1993).

[FLM]   I. B. Frenkel, J. Lepowsky, and A. Meurman. *Vertex operator algebras and the Monster.* New York: Academic Press, 1988.

[FZ]   I. B. Frenkel and Y. Zhu. Vertex operator algebras associated to representations of affine and Virasoro algebra. *Duke Math. J.*, **66** No 1 (1992), 123–168.

[FMS]   D. Friedan, E. Martinec, S. Shenker. Conformal invariance, supersymmetry and string theory. *Nucl. Phys.* **B271** (1986), 93–165.

[F]   D. B. Fuchs. *Cohomology of infinite-dimensional Lie algebras*, Contemporary Soviet Mathematics. Consultants Bureau. New York, 1986.

[FST]   P. Furlan, G. M. Sotkov, and I. T. Todorov. Two-dimensional conformal quantum field theory. *La Rivista del Nuovo Cimento*, **6** 1989.

[GF1]   I. M. Gelfand and D. B. Fuchs. *Cohomologies of the Lie algebra of vector fields on the circle*. Funkc. Anal. i Pril. **2** (1968) no. 4, 92–93.

[GF2]   I. M. Gelfand and D. B. Fuchs. *Cohomologies of the Lie algebra of formal vector fields*. Izv. Akad. Nauk SSSR Ser. Mat. **34** (1970) 322–337.

[Ge]   E. Getzler. Manin pairs and topological field theory. *Ann. Phys.* (N.Y.) **237** (1995), 161–201.

[G]   P. Goddard. Meromorphic conformal field theory. In *Infinite-dimensional Lie algebras and groups*, V. G. Kac, editor, *Adv. Ser. in Math. Phys.*, **7** (1989), World Scientific, 556–587.

[GKO]   P. Goddard, A. Kent, and D. Olive. Virasoro algebras and coset space models. *Phys. Lett.*, **B152** (1985), 88–93.

[JM]   M. Jimbo and T. Miwa. Solitons and infinite dimensional Lie algebras. *Publ. Res. Inst. Math. Sci.*, **19** (1983), 943–1001.

[K1]   V. G. Kac. Lie superalgebras. *Adv. Math.*, **26** (1977), 8–96.

[K2]   V. G. Kac. *Infinite dimensional Lie algebras*, third edition. Cambridge University Press, 1990.

[K3]   V. G. Kac. Conformal superalgebras and transitive group actions on quadrics. *Commun. Math. Phys.* **186** (1997), 233–252.

[K4]   V. G. Kac. Classification of infinite-dimensional simple linearly compact Lie superalgebras. *Adv. Math.* (1998).

[KL1]   V. G. Kac and J. van de Leur. Super boson-fermion correspondence. *Ann. Inst. Fourier* **37** (1987), 99–137.

[KL2]   V. G. Kac and J. van de Leur. On classification of superconformal algebras. In *Strings 88*, S. J. Gates et al., editors, World Scientific, 1989, 77–106.

[KL3]   V. G. Kac and J. van de Leur. The n-component KP hierarchy and representation theory. *Important developments in soliton theory*, A. S. Fokas and V. E. Zakharov, editors, Springer Verlag, 1993, 302–343.

[KRad]   V. G. Kac and A. D. Radul. Representation theory of the vertex algebra $W_{1+\infty}$. *Transformation groups 1* (1996), 41–70.

[KR]   V. G. Kac and A. K. Raina. Bombay lectures on highest weight representations of infinite-dimensional Lie algebras. *Advanced Ser. in Math. Phys.*, **2** World Scientific, 1987.

[KT1]   V. G. Kac and I. Todorov. Superconformal current algebras and their unitary representations. *Comm. Math. Phys.*, **102** (1985), 337–347.

[KT2] V. G. Kac and I. Todorov. Affine orbifolds and RCFT extensions of $W_{1+\infty}$. *Comm. Math. Phys*, **190** (1997), 57–111.

[KW1] V. G. Kac and M. Wakimoto. Modular and conformal invariance constraints in representation theory of affine algebras. *Adv. Math.*, **70** (1988), 156–234.

[KW2] V. G. Kac and M. Wakimoto. Integrable highest weight modules over affine superalgebras and number theory. In *Progress in Math.*, **123** (1994), 415–456.

[KWang] V. G. Kac and W. Wang. Vertex operator superalgebras and their representations. *Contemp. Math.*, **175** (1994), 161–191.

[KS] Y. Kazama and H. Suzuki. New $N = 2$ superconformal field theories and superstring compactification. *Nucl. Phys.*, **B321** (1989), 232–268.

[KV] T. Kimura and A. A. Voronov. The cohomology of algebras over moduli spaces. In *Progress in Math.* **129**, Birkhauser, Boston, 1995, 305–334.

[Li] H.-S. Li. Local systems of vertex operators, vertex superalgebras and modules. *J. Pure Appl. Algebra*, **109** (1996), 143–195.

[LZ] B. H. Lian and G. J. Zuckerman. Commutative quantum operator algebras. *J. Pure Appl. Algebra*, **100** (1995), 117–140.

[Man] S. Mandelstam. Dual resonance models. *Phys. Reports*, **13** (1974), 259–353.

[M] O. Mathieu. Classification des algèbras de Lie graduées simple de croissance ≤ 1. *Invent. Math.*, **86** (1986), 371–426.

[MN] A. Matsuo and K. Nagatomo. On axioms for a vertex algebra and the locality of quantum fields, preprint.

[MP] A. Meurman and M. Primc. Annihilating fields of standard modules of $gl(2, \mathbb{C})^{\sim}$ and combinatiorial identities, preprint.

[NS] A. Neveu and J. H. Schwarz. Factorizable dual model of pions. *Nucl. Phys.*, **B31** (1971), 86–112.

[OS] N. Ohta and H. Suzuki. $N = 2$ superconformal models and their free field realizations. *Nucl. Phys.*, **B332** (1990), 146–168.

[R1] J.F. Ritt. Differential groups and formal Lie theory for an infinite number of parameters. *Ann. of Math.* (2), **52** (1950), 708–726.

[R2] J.F. Ritt. Differential groups of order two. *Ann. of Math.*, **53** (1951), 491–519.

[RS] P. Ramond and J. H. Schwarz. Classification of dual model gauge algebras. *Phys. Lett.*, **64B** (1976), 75–77.

[Se] J.-P. Serre. Lie algebras and Lie groups. Benjamin, 1965.

[Sk] T. H. R. Skyrme. Kinks and the Dirac equation. *J. Math. Phys.*, **12** (1971), 1735–1743.

[S] H. Sugawara. A field theory of currents. *Phys. Rev.* **176** (1968), 2019–2025.

[T] K. Thielemans. A Mathematica package for computing operator product expansions. *Internat. Mod. Phys. C*, **2** No 3 (1991), 787–798.

[Wa] W. Wang. Rationality of Virasoro vertex operator algebras. *Duke Math. J.*, **71** No 1 (1993), 197–211.

[W] A. S. Wightman. Quantum field theory in terms of vacuum expectation values. *Phys. Rev.*, **101** (1956), 860–866.

[Wi] K. Wilson. OPE and anomalous dimensions in the Thirring model. *Phys. Rev. D*, **2** (1970), 1473–1477.

[Z] Y. Zhu. Modular invariance of characters of vertex operator algebras. *Journal AMS*, **9** (1996), 237–302.

Index